Power Transformers

POWER ENGINEERING

Series Editors

H. Lee Willis
ABB Electric Systems Technology Institute
Raleigh, North Carolina

Anthony F. Sleva
Sleva Associates
Allentown, Pennsylvania

Mohammad Shahidehpour
Illinois Institute of Technology
Chicago, Illinois

1. Power Distribution Planning Reference Book, *H. Lee Willis*
2. Transmission Network Protection: Theory and Practice, *Y. G. Paithankar*
3. Electrical Insulation in Power Systems, *N. H. Malik, A. A. Al-Arainy, and M. I. Qureshi*
4. Electrical Power Equipment Maintenance and Testing, *Paul Gill*
5. Protective Relaying: Principles and Applications, Second Edition, *J. Lewis Blackburn*
6. Understanding Electric Utilities and De-Regulation, *Lorrin Philipson and H. Lee Willis*
7. Electrical Power Cable Engineering, *William A. Thue*
8. Electric Systems, Dynamics, and Stability with Artificial Intelligence Applications, *James A. Momoh and Mohamed E. El-Hawary*
9. Insulation Coordination for Power Systems, *Andrew R. Hileman*
10. Distributed Power Generation: Planning and Evaluation, *H. Lee Willis and Walter G. Scott*
11. Electric Power System Applications of Optimization, James A. Momoh
12. Aging Power Delivery Infrastructures, *H. Lee Willis, Gregory V. Welch, and Randall R. Schrieber*
13. Restructured Electrical Power Systems: Operation, Trading, and Volatility, *Mohammad Shahidehpour and Muwaffaq Alomoush*
14. Electric Power Distribution Reliability, *Richard E. Brown*
15. Computer-Aided Power System Analysis, *Ramasamy Natarajan*

16. Power System Analysis: Short-Circuit Load Flow and Harmonics, *J. C. Das*
17. Power Transformers: Principles and Applications, *John J. Winders, Jr.*

ADDITIONAL VOLUMES IN PREPARATION

Spatial Electric Load Forecasting: Second Edition, Revised and Expanded, *H. Lee Willis*

Power Transformers
Principles and Applications

John J. Winders, Jr.
PPL Electric Utilities
Allentown, Pennsylvania

CRC Press
Taylor & Francis Group
Boca Raton London New York

CRC Press is an imprint of the
Taylor & Francis Group, an informa business

ISBN: 0-8247-0766-4

This book is printed on acid-free paper.

Headquarters
Marcel Dekker
270 Madison Avenue, New York, NY 10016
tel: 212-696-9000; fax: 212-685-4540

World Wide Web
http://www.dekker.com

The publisher offers discounts on this book when ordered in bulk quantities. For more information, write to Special Sales/Professional Marketing at the headquarters address above.

Current printing (last digit):
10 9 8 7 6 5 4 3 2

PRINTED IN THE UNITED STATES OF AMERICA

Series Introduction

Power engineering is the oldest and most traditional of the various areas within electrical engineering, yet no other facet of modern technology is currently experiencing a greater transformation or seeing more attention and interest from the public and government. But while public concern and political decisions about de-regulation and energy trading may reshape the electric utility industry's manner of doing business, its future like its past rests on the capability of its transmission and distribution systems to convey safe, reliable, and economical electric power to homes, businesses, and factories. Nothing is more essential to this performance than the transformer, which enables modern power and industrial systems to function.

I am particularly delighted to see this latest addition to Marcel Dekker's Power Engineering series. *Power Transformers: Principles and Applications* is a comprehensive compendium of theory and practices for electric power transformers. This book provides a concise but thorough treatment of basic transformer theory, its application to various types of transformer designs and their application in utility and industrial power systems. Its easy to read style and linear organization make it particularly suitable as a tutorial for those who need to learn the material independently, outside of the classroom, or as a text for formal courses. This book also makes a very good practical reference for utility and industrial power engineers.

In addition to having concise summaries of all the basics, the text provides an excellent description of the various ancillary equipment and systems, which are often the most difficult to precisely engineer and fit into the system. John Winders has also provided excellent coverage of how to read, interpret, and apply a power transformer's nameplate data, not always a straightforward or unambiguous task and one where a surprising number of mistakes are made by inexperienced engineers. Chapter 8 will be particularly useful to practicing engineers and power system operators, covering maintenance needs, testing options, and troubleshooting techniques and their use, and discussing reliability of transformers.

As the editor of the Power Engineering series, I am proud to include *Power Transformers: Principles and Applications* among this important group of books. Like all the books in Marcel Dekker's Power Engineering series, this book provides modern power technology in a context of proven, practical application, useful as a reference book as well as for self-study and advanced classroom use. Marcel Dekker's Power Engineering series includes books covering the entire field of power engineering, in all of its specialties and subgenres, each aimed at providing practicing power engineers with the knowledge and techniques they need to meet the electric industry's challenges in the 21st century.

H. Lee Willis

Preface

This book is based on notes for the Transformer Applications Course offered by the Center for Power System Study at Lehigh University. The key word in both the title of that course and the title of this book is *applications*. The material presented in the following chapters was obtained from various sources: textbooks, industry standards, and established utility practices and procedures. Much of this material also comes from my personal files relating to actual events and case studies that were observed during my career in the utility industry spanning 30 years.

There are many kinds of transformers, and all share the same set of fundamental operating principles. Since this book focuses on *power* transformers, it is fair to ask, "What exactly *is* a power transformer?" By definition, a power transformer is a transformer which transfers electric energy in any part of the circuit between the generator and the distribution primary circuits.* This definition of power transformer in the IEEE standard appears under the

* IEEE Std. C57.12.80-1978. IEEE Standard Terminology for Power and Distribution Transformers. Institute of Electrical and Electronics Engineers, Inc., 1978, New York, p. 8.

heading of ''Size'' and does not indicate how the transformer is used in the power system. Thus, this book uses this definition in the broadest sense to include discussions of specialty applications such as step voltage regulators, phase shifters, and grounding transformers, as well as the usual step-up and step-down applications. Since the line between power transformers and distribution transformers is somewhat blurry, many of the basic principles presented can be applied to distribution transformers as well.

The first several chapters build a solid theoretical foundation by describing the underlying physics behind transformer operation. A theoretical foundation is absolutely necessary in order to understand what is going on inside a transformer and why. The magnetic properties of materials, a review of magnetic units, and analysis of magnetic circuits are discussed with enough mathematical rigor for the interested reader to gain full comprehension of the physics involved. Whenever a detailed mathematical treatment is presented, it is always done with a practical objective in mind. Each chapter includes a number of practice problems to clearly illustrate how the theory is applied in everyday situations. Many of these practice problems are based on actual events.

Several things set this book apart from other transformer reference books. First, this book emphasizes the importance of magnetic properties and how the choice of a core design can affect the transformer's electrical properties, especially during faults and unbalanced operations. Many reference books overlook this critical aspect of transformer applications.

Next, this book discusses special types of transformer connections, such as the zigzag, Scott, and tee connections, as well as the more common wye and delta types. The Scott and tee connections, which transform three-phase voltages into two-phase voltages, are seldom covered in modern transformer reference books even though two-phase systems still exist today. Tap changing under load and variable phase shifting transformers are covered. Different types of transformer coil and coil construction are compared, with discussion of the particular advantages and disadvantages of each with respect to the various transformer connections. The reader will also gain insight into some of the economic trade-offs of different transformer design options.

A brief tutorial on symmetrical components is also included. The topic is covered in other reference books but seldom in such a compact and straightforward way, enabling the reader to immediately apply the technique in practical problems.

A section of the book defines a transformer's nameplate rating versus its thermal capability and describes how to calculate a transformer's rate of loss of life. An entire chapter is devoted to describing abnormal operating conditions that can damage power transformers, including overloads, short circuits, single phasing from primary fuse operations, ferroresonance, and voltage surges. The chapter describes ways to avoid these conditions, or at least ways to mitigate them through proper system design and selection of appropriate transformer designs.

The reader will learn how to interpret and use a transformer test report as well as the information on the transformer nameplate. The book concludes with a comprehensive discussion of preventive and predictive maintenance, good utility practices, factory and field testing, and failure rate analysis.

This book is intended primarily for readers having an electrical engineering background although training as an electrical engineer is not necessary, and others will also benefit from the conclusions that can be drawn from the practical examples. Mastery of the principles presented in this book will provide a sound working knowledge of how to specify, operate, and maintain power transformers in a utility or plant environment.

I wish to thank Anthony F. Sleva for his thorough review of the manuscript and his many helpful suggestions for improving it, and for making it possible to publish this book. I am indebted to the late Charles H. Morrison, who patiently shared with me so much of his extensive theoretical and practical knowledge about power transformers.

John J. Winders, Jr.

Contents

Series Introduction H. Lee Willis *iii*
Preface *v*

1 Basic Transformer Theory **1**

1.1 Definition of a Transformer 1
1.2 Magnetic Units and Conversion Factors 1
1.3 Currents and Magnetic Fields 3
1.4 Magnetic Induction 4
1.5 Constructing a Simple Transformer 5
1.6 The Magnetic Circuit 6
1.7 The *B-H* Curve 10
1.8 The *B-H* Curve and Hysteresis 11
1.9 Magnetizing Currents and Harmonics 15
1.10 Transformer Core Design and Construction 16
1.11 Magnetostriction 20
1.12 Completing the Transformer by Adding a Second Winding 20

1.13 Placement of the Windings 22
References 24

2 **Two-Winding Transformer Connections** **25**

2.1 Introduction 25
2.2 The Y-Y Connection in Three-Phase Systems 26
2.3 Advantages of the Y-Y Connection 29
2.4 Disadvantages of the Y-Y Connection 30
2.5 The Y-Δ Connection and the Δ-Y Connection 33
2.6 Phase Angle Displacement and Phase Rotation 36
2.7 The Y-Δ Grounding Bank 40
2.8 The Zigzag Connection 42
2.9 Comparisons of Economy of the Different Winding Configurations 44
2.10 Trade-Off Between Steel and Copper in the Design of a Transformer 47
2.11 Connecting Three-Phase Banks Using Single-Phase Transformers 48
2.12 Transforming Three-Phase Voltages into Two-Phase Voltages 50
2.13 The Scott Transformer Connection 51
2.14 Three-Phase Transformer Designs 56
2.15 Standard Terminal Markings for Transformers 62
References 67

3 **Transformer Impedance and Losses** **69**

3.1 Leakage Flux and Leakage Reactance 69
3.2 Conductor Losses 74
3.3 No-Load Losses 76
3.4 Magnetizing Reactance 79
3.5 Equivalent Circuit of a Two-Winding Transformer 80
3.6 A Brief Tutorial on Symmetrical Components 81
3.7 Transformer Equivalent Circuits Modeled in Zero Phase Sequence Networks 85
3.8 Series Impedance and Regulation 92
3.9 Matching Transformers for Parallel and Bank Operations 93

3.10 Impedance Mismatch in Three-Phase Transformer Banks 95
3.11 Temperature Rise and the Thermal Capability 98
3.12 Interpreting Transformer Test Reports 99
3.13 Calculating the Hot-Spot Temperature Using the IEEE Method 102
3.14 Calculating the Loss of Life 105
References 116

4 Autotransformers and Three-Winding Transformers 117

4.1 Autotransformer Connections 117
4.2 Impedance of an Autotransformer 122
4.3 Limitations of the Autotransformer Connection 128
4.4 Autotransformer Voltages with Short Circuits Applied 129
4.5 Impulse Voltages Applied to Autotransformers 131
4.6 Autotransformer Core and Coil Designs and Terminal Configurations 131
4.7 Advantages and Disadvantages of the Autotransformer Connection 132
4.8 Three-Winding Transformers 133
4.9 Modification of Transformer Laws with Three Windings 134
4.10 Equivalent Circuit of a Three-Winding Transformer 136
4.11 Core and Coil Construction of Three-Winding Transformers 138
4.12 Thermal Capability of Three-Winding Transformers 141
4.13 The Stabilizing Effect of a Δ Tertiary Winding 141
Reference 146

5 Short Circuits, Inrush Currents, and Other Phenomena 147

5.1 Effects of Short Circuits on Transformers 147
5.2 Comparisons of Short-Circuit Currents for Various Faults 148
5.3 Mechanical Forces in Transformers 155
5.4 Forces between Transformer Windings 158
5.5 Short-Circuit Forces in Three-Winding Transformers 163
5.6 Exciting Current Inrush 165
5.7 Tank Overheating from Zero-Sequence Currents 169

5.8 Primary Fuse Misoperations 172
5.9 Ferroresonance 175
5.10 Voltage Surges 177
References 187

6 Ancillary Equipment 189

6.1 Bushings 189
6.2 Oil Preservation Systems 192
6.3 Cooling Equipment 196
6.4 Temperature, Oil Level, and Pressure Gauges 201
6.5 Tap-Changing Equipment 202
6.6 Load Tap Changers 204
6.7 Voltage Regulating Transformers 209
6.8 Tap Changer Automatic Controls 210
6.9 Variable Phase Shifting Transformers 212
Reference 220

7 Reading and Applying Nameplate Information 221

7.1 Minimum Nameplate Requirements 221
7.2 Manufacturer's Information 223
7.3 Cooling Class, Number of Phases, and Operating Frequency 223
7.4 Voltage Ratings 224
7.5 KVA or MVA Ratings 225
7.6 Winding Connection Diagram 226
7.7 Phasor or Vector Diagram 229
7.8 Weights and Oil Capacity 229
7.9 Operating Pressure Range 230
7.10 Impedance 231
7.11 Basic Insulation Level 232
7.12 Nameplate Layout 233
References 233

8 Maintenance, Testing, Troubleshooting, and Reliability 235

8.1 Good Utility Practices 235
8.2 Preventative Maintenance versus Predictive Maintenance 252

8.3	Factory Tests	252
8.4	Ratio Test	252
8.5	Other Factory Tests	255
8.6	Field Tests	258
8.7	Gas-in-Oil Analysis	261
8.8	Water-in-Oil Analysis	265
8.9	Drying Transformers	268
8.10	Oil Dielectric Test	268
8.11	Reliability Calculations	269
8.12	The Chi-Squared Distribution	269
8.13	The Poisson Distribution	271
8.14	Statistical Economics	273
	References	275

Index *277*

Power Transformers

1

Basic Transformer Theory

1.1 DEFINITION OF A TRANSFORMER

A transformer is "a static device consisting of a winding, or two or more coupled windings, with or without a magnetic core, for inducing mutual coupling between circuits. Note: Transformers are exclusively used in electric power systems to transfer power by electromagnetic induction between circuits at the same frequency, usually with changed values of voltage and current." [1]

There are numerous types of transformers used in various applications including audio, radio, instrument, and power. This book deals exclusively with power transformer applications involving the transmission and distribution of electrical power. Power transformers are used extensively by traditional electric utility companies, power plants, and industrial plants.

1.2 MAGNETIC UNITS AND CONVERSION FACTORS

The basic operation of all transformers is deeply rooted in electromagnetics, whether or not the transformer has a magnetic iron core. Students are often confused by the terminology used to describe magnetic phenomena. Part of the confusion lies in the different units of measurement that are used. There are three basic systems of measurement used in engineering: English, MKS (meter-kilogram-second), and cgs (centimeter-gram-second). To make matters worse, some transformer textbooks even mix English units with cgs or MKS units. For consistency and ease of understanding, this book will use MKS units throughout the example problems.

The fist magnetic quantity is the magnetomotive force (MMF). In electrical terms, MMF is roughly equivalent to the electromotive force (EMF), that causes current to flow in an electrical circuit. The units and conversion factors for MMF are

MKS: ampere-turn
cgs: gilbert
1 Gb = 0.4π amp-turn

The next magnetic quantity is flux, represented by the Greek letter ϕ. Since a magnetic field can be visualized as a bundle of lines flowing from a north magnetic pole to a south magnetic pole, flux is the total number of "lines." The units and conversion factors of ϕ are

MKS: weber
cgs: maxwell
1 Ma = 1 line = 10^{-8} Wb

The magnetic flux density B is the concentration of magnetic of lines across an area. The units and conversion factors for B are

MKS: tesla
cgs: gauss
1 G = 10^{-4} T
1 T = 1 Wb/m^2

The magnetic field intensity H is the distribution of MMF along a magnetic path. If the flux density is constant, H is merely the total MMF divided by the length of the magnetic path. The units and conversion factors for H are

MKS: amp-turns/meter
cgs: oersted
1 Oe = $(250/\pi)$ amp-turns/m

1.3 CURRENTS AND MAGNETIC FIELDS

Consider the straight cylindrical conductor carrying a current i shown in Figure 1.1 A magnetic field surrounds the conductor. According to the right-hand rule, a magnetic field surrounds the conductor in a counterclockwise direction. The right-hand rule is stated as follows: With the thumb of the right hand pointing in the direction of the electrical current, the fingers point in the direction of the magnetic field. When applying the right-hand rule, it is important to use *conventional* electrical current and not the *electron current*.

For any closed path around the conductor with the incremental length dl, in the direction of the magnetic field, the magnetic flux density, B is a function of the current in the conductor according to the following equation:

$$\int B \cdot dl = i \times \mu_0 \tag{1.3.1}$$

where μ_0 is the vacuum permeability $= 4\pi \times 10^{-7}$ N/A^2.

For a straight conductor, the path of B around the conductor is always circular, so at a distance r from the center of the conductor, the integral in Eq. (1.3.1) is reduced to

$$\int B \cdot dl = B \times 2\pi r \tag{1.3.2}$$

Therefore, the magnetic field intensity is inversely proportional to the distance from the center of the conductor

$$B = i \times \frac{\mu_0}{2\pi r} \tag{1.3.3}$$

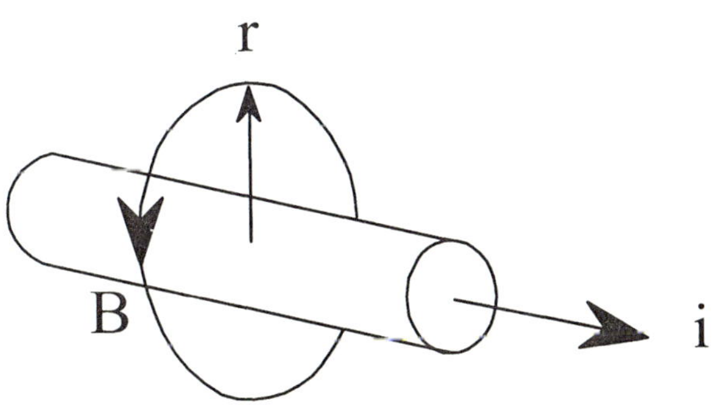

FIGURE 1.1 *B* field surrounding a straight cylindrical conductor carrying a current *i*.

Example 1.1

A straight conductor carries a current of 100 A. What is the magnetic field intensity at a distance of 25 cm from the center of the conductor?

In the MKS system, $r = 0.25$ m. Using Eq. (1.3.3),

$$B = 100 \text{ A} \times 4\pi \times 10^{-7} \text{ N/A}^2/(2\pi \times 0.25 \text{ m}) = 8 \times 10^{-5} \text{ T}$$

1.4 MAGNETIC INDUCTION

For a closed path in a magnetic field, the total flux ϕ is found by integrating the incremental surface area dA times the normal component of the magnetic field intensity B over any surface within the closed path:

$$\phi = \int B \cdot dA \tag{1.4.1}$$

where

ϕ = flux, Wb
dA = incremental surface area, m^2

If the total flux is changing over time, there is an induced voltage E around the closed path surrounding ϕ. The value E in volts is equal to $-d\phi/dt$, where the direction of E is in the right-hand sense. Figure 1.2 illustrates this principle of magnetic induction. If the magnitude of B is decreasing, then $d\phi/dt$ will be in the downward direction, and E will be in the positive in the right-hand sense around the closed loop that encircles ϕ.

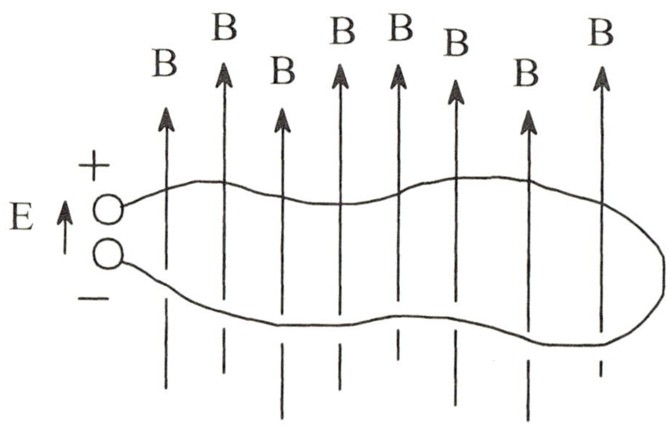

FIGURE 1.2 Voltage induced in a loop surrounding a time-varying B field.

1.5 CONSTRUCTING A SIMPLE TRANSFORMER

From the foregoing discussion of the basic principles of magnetic induction, it is not difficult to see how a rudimentary transformer could be constructed. If a conductor carrying a changing current is brought near a second conductor, then the changing magnetic flux surrounding the first conductor will be linked to the second conductor and will induce a voltage. Such a rudimentary transformer is depicted in Figure 1.3.

An AC voltage is connected to a *primary* conductor, shown as the left-hand solid bar in Figure 1.3. In response to the voltage, an AC current flows, setting up a time-varying magnetic field surrounding the primary conductor. A *secondary* conductor, shown as the right-hand solid bar, is located in proximity to the primary conductor so that the magnetic flux surrounding the primary conductor *links* the secondary circuit. According to the law of induction, there will be an induced voltage E around the path surrounding the time-varying flux.

The configuration shown above is not very efficient in transferring energy because only a small portion of the total magnetic flux surrounding the primary conductor will be linked to the secondary circuit. In order to improve the efficiency of the rudimentary transformer, the magnetic field needs to be channeled in such a way that most of the flux produced by the primary conductor is linked to the secondary circuit. This is accomplished by surrounding the primary and secondary conductors with a magnetic core material having an affinity for magnetic flux. This modification is shown in Figure 1.4. By adding the magnetic core, essentially all of the magnetic flux produced in the primary

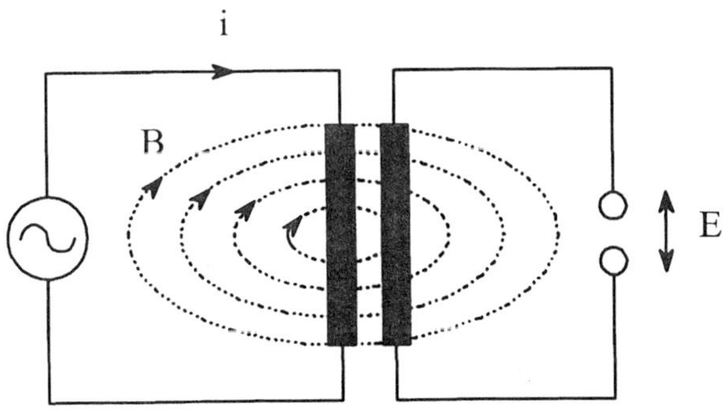

FIGURE 1.3 Voltage induced in a conductor.

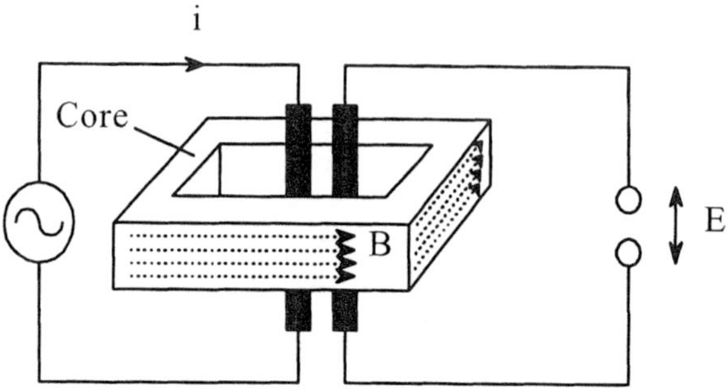

FIGURE 1.4 Channeling a B field through a magnetic core.

conductor is linked to the secondary conductor. Therefore, the efficiency of the rudimentary transformer is greatly increased.

Various types of core materials exist. The important physical property is the *permeability constant* μ, given in units of N/A^2. The *relative permeability* μ_r is the permeability constant divided by the vacuum permeability μ_0. Values of μ_r for some common magnetic core materials are as follows:

SiFe (unoriented)	400
SiFe (oriented)	1500
50–50 NiFe (oriented)	2000
79 Permaloy	12,000–100,000

A grain-oriented silicon steel conducts magnetic flux 1500 times better than a vacuum. The advantages and disadvantages of grain-oriented steels will be discussed in a later chapter.

The ratio of the flux density B and the field intensity H is equal to the permeability of the medium μ:

$$\mu = \frac{B}{H} \tag{1.5.1}$$

$$H = \frac{B}{\mu} \tag{1.5.2}$$

1.6 THE MAGNETIC CIRCUIT

Since the magnetic core has been introduced, an understanding of the magnetic circuit is necessary to quantify the relationships between voltage, current, flux, and field density.

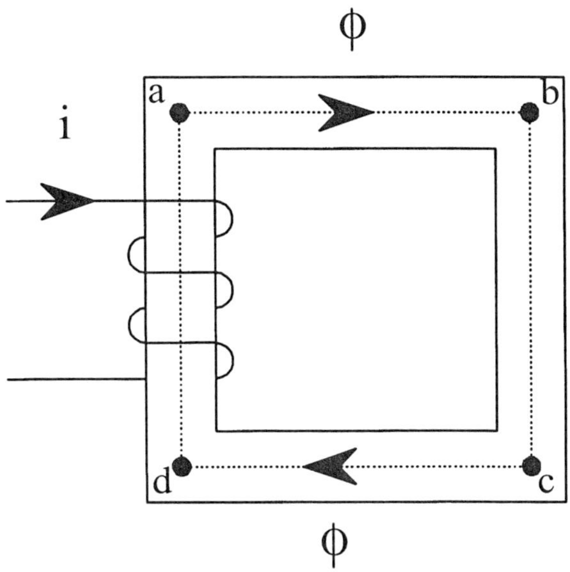

FIGURE 1.5 Closed magnetic circuit.

Consider the magnetic circuit shown in Figure 1.5 consisting of a coil of wire wound around a magnetic yoke. The coil has N turns and carries a current i. The current in the coil causes a magnetic flux to flow along the path *a-b-c-d-a*. For the time being, let us assume that the flux density is small so that the permeability of the yoke is a constant. The magnitude of the flux is given by

$$\phi = N \times i/\mathcal{R} \tag{1.6.1}$$

where $N \times i$ is the magnetomotive force (MMF) in ampere-turns and $\mathcal{R}$ is the *reluctance* of the magnetic circuit *a-b-c-d-a*.

As the name implies, reluctance is a property that resists magnetic flux when MMF is applied to a magnetic circuit. Reluctance is roughly equivalent to the resistance in an electrical circuit.

For a homogeneous material where the mean length of the flux path is l and the cross-sectional area is A, the reluctance is calculated in the MKS system of measurement as follows:

$$\mathcal{R} = \frac{l}{\mu \times A} \qquad A^2/J \tag{1.6.2}$$

The coil's inductance L is equal to $N^2(\mu \times A)/l$. Therefore, the coil's inductance is inversely proportional to reluctance of the magnetic circuit. For series elements in the magnetic path, the total reluctance is found by adding the values of reluctance of the individual segments along the magnetic path. The reluctance values of parallel elements in a magnetic circuit are combined in a manner similar to combining parallel resistances in an electrical circuit.

Example 1.2

In the magnetic circuit shown in Figure 1.6, the coil has 100 turns and carries 10 A. The relative permeability of the yoke is 10,000. The lengths of the segments along the mean magnetic path are as follows:

$$\text{Segments } a\text{-}b \text{ and } e\text{-}f = 10 \text{ cm}$$
$$\text{Segments } b\text{-}c \text{ and } d\text{-}e = 4 \text{ cm}$$
$$\text{Segment } f\text{-}a = 9 \text{ cm}$$
$$\text{Air gap} = 1 \text{ cm}$$

The cross-sectional area of all segments is 4 cm^2. Find the flux ϕ and the flux density B.

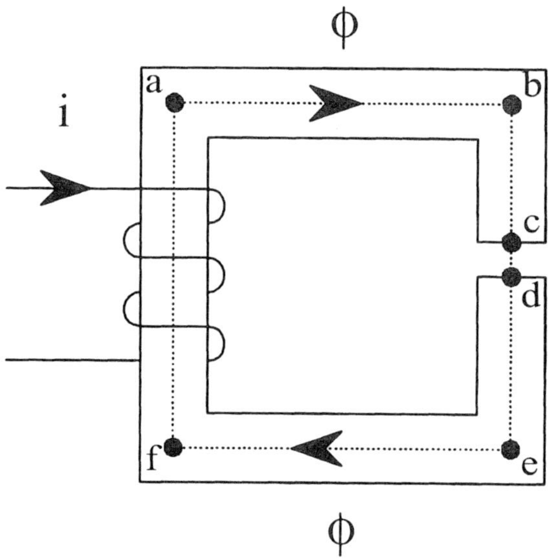

FIGURE 1.6 Magnetic circuit with an air gap.

Using the MKS system of measurement, the reluctance of the yoke is $\mathcal{R}_y$:

$$\mathcal{R}_y = \frac{2 \times 0.10 \text{ m} + 2 \times 0.04 \text{ m} + 0.09 \text{ m}}{10{,}000\mu_0 \times 0.0004 \text{ m}^2} = \frac{0.0925 \text{ m}^{-1}}{\mu_0}$$

The reluctance of the air gap is $\mathcal{R}_a$:

$$\mathcal{R}_a = \frac{0.01 \text{ m}}{\mu_0 \times 0.0004 \text{ m}^2} = \frac{25 \text{ m}^{-1}}{\mu_0}$$

The total reluctance $\mathcal{R}_T$ is found by adding $\mathcal{R}_y$ and $\mathcal{R}_a$:

$$\mathcal{R}_T = \mathcal{R}_y + R_a = \frac{25.0925 \text{ m}^{-1}}{\mu_0} = \frac{25.0925}{4\pi \times 10^{-7}} \text{ A}^2/\text{J}$$

The flux is found by dividing the MMF by the total reluctance:

$$\phi = N \times \frac{i}{\mathcal{R}_t} = 100 \times 10 \times 4\pi \times \frac{10^{-7}}{25.0925} \text{ Wb} = 5.008 \times 10^{-5} \text{ Wb}$$

The flux density is found by dividing the flux by the cross-sectional area of the magnetic path:

$$B = \frac{\phi}{A} = 5.008 \times \frac{10^{-5} \text{ Wb}}{0.0004 \text{ M}^2} = 0.1252 \text{ Wb/M}^2$$

The magnetic field intensity H is equal to the flux density B divided by the permeability:

$$H = \frac{B}{\mu}$$

In the yoke,

$$\begin{aligned} H &= \frac{0.1252}{10{,}000 \times \mu_0} \text{ Wb/m}^2 \\ &= \frac{0.1252}{(10{,}000 \times 4\pi \times 10^{-7})} \text{ amp-turn/m} \\ &= 9.963 \text{ amp-turns/m} \end{aligned}$$

In the air gap, $\mu = \mu_0$:

$$H = \frac{0.1252}{4\pi \times 10^{-7}} \text{ amp-turns/m} = 9.963 \times 10^4 \text{ amp-turns/m}$$

The magnetic field in the air gap sets up an attractive force that tends to pull the pole pieces of the yoke together. The force F in the MKS system of measurement is given by

$$F = \frac{B^2 A}{2\mu_0} \text{ N} \tag{1.6.3}$$

For the magnetic circuit in this example the mechanical force across the air gap is calculated from Eq. (1.6.3) as follows:

$$F = \left(\frac{0.1252 \text{ Wb}}{\text{m}^2}\right)^2 \times \frac{0.0004 \text{ m}^2}{2 \times 4\pi \times 10^{-7} \text{ A}^2/\text{J}} = 2.49 \text{ N}$$

1.7 THE *B-H* CURVE

Up to this point, it was assumed that the core permeability is constant; i.e., $B = \mu \times H$. For actual transformer core materials, the relationship between B and H is much more complicated. For a flux that periodically changes, the B-H curve depends on the magnitude of the flux density and the periodic frequency. Figure 1.7 plots the B-H curve for a ferromagnetic core with a 60 Hz sinusoidal flux density having a moderate peak value.

The B-H curve is a closed loop with the path over time moving in a counterclockwise direction over each full cycle. Note that when the magnetizing current is zero ($H = 0$) there is still a considerable positive or negative residual flux in the core. This residual flux is from crystalline structures in ferromagnetic materials that remain magnetically aligned even after the MMF is removed.

For a given peak amplitude of flux density, the B-H loop becomes narrower at frequencies below 60 Hz, although the width of the loop is not directly proportional to frequency. Even at very low frequencies approaching DC, the B-H curve has a finite area contained in the loop.

As seen in Figure 1.7, magnetic materials are highly nonlinear, so treating m as a constant is clearly an oversimplification. Nevertheless, assuming that materials are linear, at least over some range of flux density, is required in order to do quantitative analysis.

As the peak amplitude of the flux increases, the core goes into *saturation*; i.e., B increases at a much smaller rate with respect to increasing H. This means that μ gets effectively smaller as B increases. In saturation, the slope dB/dH is approximately equal to μ_0. Figure 1.8 plots a typical B-H curve for a ferromagnetic core with a 60 Hz sinusoidal flux density having a large peak

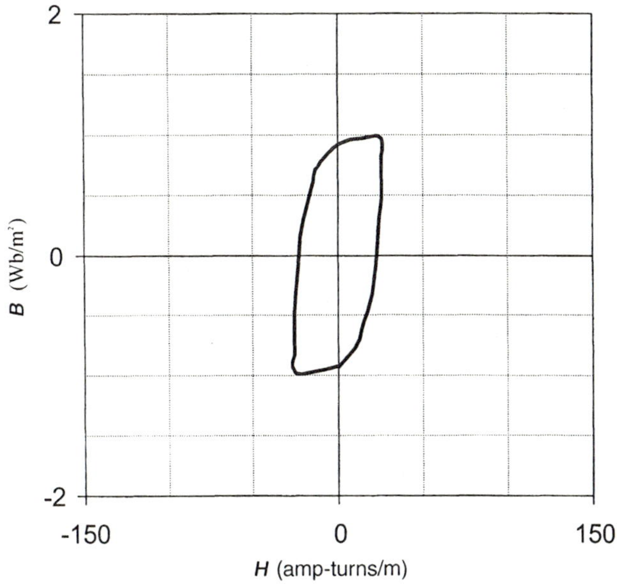

FIGURE 1.7 *B-H* curve for moderate flux density.

value. This core material saturates at approximately ± 1.5 Wb/m^2 (± 1.5 T), which is a typical saturation value for materials used in power transformers. The magnitude of H increases greatly when the core goes into saturation, meaning that the peak magnetizing current increases dramatically. Again, the width of the *B-H* loop becomes narrower at frequencies below 60 Hz for a given peak amplitude of flux.

1.8 THE *B-H* CURVE AND HYSTERESIS

Suppose coil having N turns of wire is wound around a magnetic core and the coil conducts a time-varying current i. The current magnetizes the core, inducing a voltage across each turn. The quantity of volts per turn in the MKS system of measurement is given by

$$E/N = -\frac{d\phi}{dt} \tag{1.8.1}$$

If ϕ is expressed in cgs units, then the volts per turn will be 10^{-8} times this value. Note that the induced voltage per turn is the same for *any* turn of

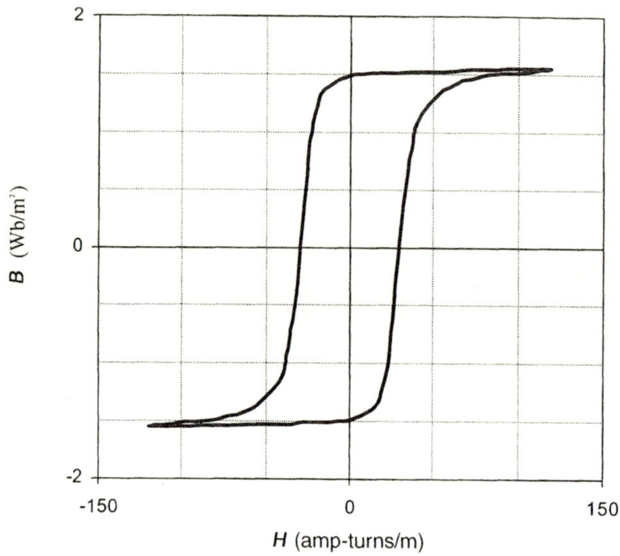

FIGURE 1.8 *B-H* curve for large flux density.

wire wound around core, including the coil carrying the magnetizing current. The voltage across the entire coil E is simply the volts per turn times the number of turns.

$$E = -N \times \frac{d\phi}{dt} \qquad (1.8.2)$$

The minus sign indicates that the induced voltage tends to oppose the direction of the current flow. The energy supplied to the coil from the electrical circuit W is found by integrating the magnitude of the power supplied to the coil over time. Since the power is equal to the voltage across the coil times the current in the coil, and ignoring the electrical resistance of the coil,

$$W = \int E \times i\, dt = \int N \times \frac{d\phi}{dt} \times i\, dt = \int N \times i\, d\phi \quad \text{J} \qquad (1.8.3)$$

Suppose the core has a mean length equal to l, and assume the core has a uniform cross-sectional area equal to A.

$$N \times i = H \times l \qquad (1.8.4)$$

$$d\phi = A \times dB \tag{1.8.5}$$

$$\text{Core volume in meter}^3 = A \times l \tag{1.8.6}$$

$$W = (\text{core volume in meter}^3) \times \int \mathrm{H}\, dB \quad \mathrm{J} \tag{1.8.7}$$

$$\int H\, dB = \text{area contained in the } B\text{-}H \text{ loop} \tag{1.8.8}$$

In the MKS system with a periodically changing current, the energy delivered to the coil over each complete cycle is equal to the physical volume of the core material times the area contained in the *B-H* loop. The energy supplied to the coil is called *hysteresis loss*, which dissipates as heat in the core. It results from a kind of "friction" that occurs when the magnetic domains of the core material realign every half-cycle.

Example 1.3

A magnetic core has a uniform cross-sectional area and a total volume of 1 m^3. Given the graph of a 60 Hz *B-H* curve, as shown in Figure 1.9, estimate the total hysteresis losses in watts for this core.

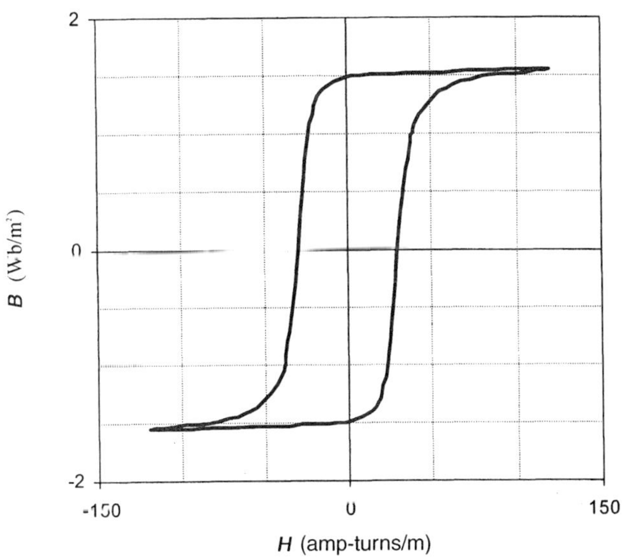

FIGURE 1.9 *B-H* curve used in Example 1–4.

Each of the dashed rectangles shown in Figure 1.9 represents $0.5 \times 50 = 25$ J of energy per cubic meter of core material. The hysteresis loop occupies roughly 9 rectangles equaling 225 J per cycle per cubic meter. At 60 Hz, there are 13,500 J dissipated per cubic meter every second, so the total hysteresis losses are about 13.5 kW for a 1 m^3 core.

Example 1.4

Suppose a coil having 100 turns is wound on a core with a uniform cross-sectional area of 0.25 m^2 and a mean path length of 4 m. Using the 60 Hz hysteresis curve shown in Figure 1.9, what is the sinusoidal voltage that is required to excite the core to the level shown in the figure, and what is the peak magnetizing current?

From the *B-H* curve in Figure 1.9, the peak value of *H* is around 125 amp-turns/m and the peak value of *B* is around 1.5 Wb/m^2. The MMF is found by multiplying *H* by the mean core length:

$$\text{MMF} = 125 \text{ amp-turns m} \times 4 \text{ m} = 500 \text{ amp-turns}$$
$$i = \text{MMF}/N = 500 \text{ amp-turns}/100 \text{ turns} = 5 \text{ A (peak)}$$

The flux is found by multiplying *B* by the cross-sectional area of the core:

$$\phi = B \times A = 1.5 \times 0.25 \text{ Wb} = 0.375 \text{ Wb (peak)}$$

For a frequency of 60 Hz,

$$\phi = 0.375 \sin(377t) \quad \text{Wb}$$
$$E = -N\, d\phi/dt = -100 \times 0.375 \times 377 \cos(377t) \quad \text{V}$$

The root mean square (RMS) voltage is found by multiplying the magnitude of *E* times 0.707:

$$E_{\text{RMS}} = 100 \times 0.375 \times 377 \times 0.707 = 9995 \text{ V}$$

The apparent power supplied to the coil is the RMS voltage times the RMS current. If the magnetizing current were purely sinusoidal, then $i_{\text{RMS}} = 5 \times 0.707$ A $= 3.535$ A and the apparent power would be $9995 \times 3.535 = 35.332$ KVA. Remembering that the hysteresis losses found in Example 1.3 were 13.5 kW, the power factor of the load supplied to the coil is around 38%. In reality, however, the magnetizing current cannot be assumed to be purely sinusoidal, as will be seen in the next section.

1.9 MAGNETIZING CURRENTS AND HARMONICS

The B-H curve can be used to construct the wave shape of the magnetizing current from the wave shape of the excitation voltage. Figure 1.10 illustrates the technique. First, the integral of the exciting voltage divided by $N \times A$ is plotted along the horizontal axis. In this case, the exciting voltage is assumed to be sinusoidal so its integral is cosinusoidal. The integral of E is divided by $(N \times A)$ to obtain B.

Each point on the B cosine curve is projected onto a point on a curve along the vertical axis, using the B-H curve to determine the horizontal distance from the vertical axis. Some of these projections are indicated by the dotted lines below. The value of H corresponding to this point is multiplied by the mean length of the core to obtain the magnetizing current i. It is clearly seen from this example that that the magnetizing current is not sinusoidal.

A Fourier analysis of magnetizing current shows a 60 Hz fundamental and the presence of odd harmonics. Generally speaking, the harmonic content of the magnetizing current increases as the level of excitation increases, especially as the core goes into saturation. For a moderate flux density of 12 KG, the percent harmonic content of the exiting current is shown in Table 1.1.

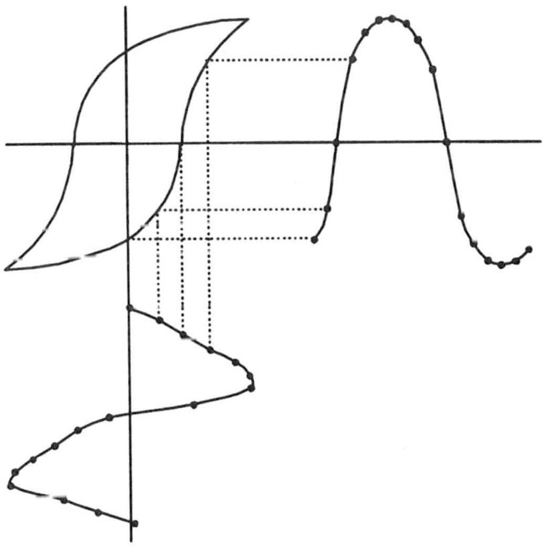

FIGURE 1.10 Developing the plot of magnetizing current from a B-H curve.

Table 1.1 Harmonic Content of the Exciting Current for a Moderate Level of Excitation

Harmonic	3rd	5th	7th	9th	11th
Content present in exciting current, %	20	6	2	0.7	0.3

Source. Ref. 2.

The predominant harmonic is the third (180 Hz), although the higher-order harmonics are strong enough to cause telephone interference problems in circuits in proximity to power lines. The presence of third harmonics in an electric power system can also cause equipment to overheat and produce voltage distortion. The isolation of third harmonics is an important design consideration in transformers. Techniques to accomplish this are covered in later chapters.

1.10 TRANSFORMER CORE DESIGN AND CONSTRUCTION

As we have seen in Example 1.2, air gaps in a magnetic core will add considerable reluctance to the magnetic circuit. Remembering that the inductance of a coil and the magnetic reluctance are inversely proportional, air gaps reduce the inductance of the coil and increase the magnitude of magnetizing currents. In practical transformers, we want to reduce magnetizing currents to almost negligible levels; it is therefore important to eliminate all air gaps if possible. One approach would be to make the core from a solid block of material. This is impractical from the standpoint of fabricating the transformer, since the coils would have to be wound through the core window.

Also, since metallic core materials conduct electric current as well as magnetic flux, the induced voltages would produce large circulating currents in a solid core. The circulating currents would oppose the changing flux and effectively "short out" the transformer.

A practical solution is to fabricate the core from thin laminated steel sheets that are stacked together and to coat the surfaces of the laminations with a thin film that electrically insulates the sheets from each other. Steel not only has excellent magnetic properties but is also relatively inexpensive and easy to fabricate into thin sheets.

In a modern transformer plant, steel ribbon is cut into sections by a cutting/punching machine commonly called a Georg machine. The sizes and shapes of the sections are determined by the core design of the individual transformer. The thickness of the sheets varies somewhat; core laminations operating at 60 Hz are between 0.010 and 0.020 in. thick, with 0.012 in. being the most common thickness in use today.

Different methods of stacking core steel have been used in the past. One such method is called the butt lap method using rectangular core sections and is illustrated in Figure 1.11 Even if the edges of the segments do not butt together perfectly, as shown in the exaggerated edge view at the bottom of the figure, the alternating even and odd layers assure that the magnetic flux has a continuous path across the surfaces of the adjacent layers. The type of construction depicted above works best with core steel that is not *grain oriente*, i.e., where permeability does not depend on the direction of the flux through the steel.

One of the greatest contributions to transformer efficiency and low cost was the introduction of grain-oriented steel in the 1940s. Grain-oriented steel is a silicon-iron alloy that is rolled or "worked" during fabrication in such a way that the permeability is higher and the hysteresis losses are lower when the flux is in the direction of the "grain." Unfortunately, the properties of this steel for a flux that goes "against the grain" are much worse than the non-grain-oriented steel. Therefore, the design of the core has to take this into account. When using grain-oriented steel, the lamination sections are mitered at a 45° angle so that when the flux changes direction by 90°, it more or less

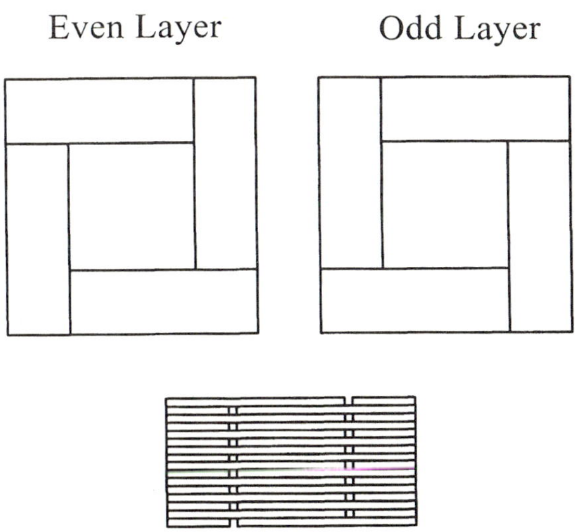

FIGURE 1.11 Stacking a laminated core using butt lap construction.

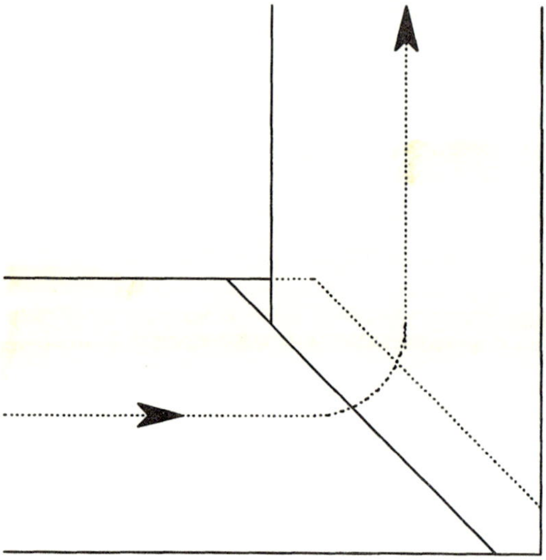

FIGURE 1.12 Flux transition at the corner of a mitered core.

still follows the grain of each of the segments. Figure 1.12 depicts a corner overlap using core sections with 45° mitered edges. The grain of the steel is oriented along the length of the laminations in the horizontal and vertical directions. The flux is at a 45° angle to the grain at the mitered edges. Alternate layers are cut into slightly different lengths and their corners have slightly different shapes. The modern multistep layer method uses up to five layers of differently shaped sections. This method is illustrated in Figure 1.13 with the first three layers exploded to show the stacking sequence.

The cross section of a transformer core can either be square or rectangular; however, a round shape is used in most large transformers of the so-called *core form* design, where the coils have a round cross section. With a round core within round coils, the use of space and materials is more efficient. Attaining a round cross section with thin steel laminations is not that difficult, although this complicates the design. The design engineer is usually content with approximating a circular core, as shown schematically in Figure 1.14. Note that the actual laminations are too thin to show individually. The empty spaces between the core and the circular coil are filled with wooden dowels or other spacer materials to improve the mechanical strength of the transformer. Chapter 5 will include a discussion of the mechanical forces that occur under fault conditions.

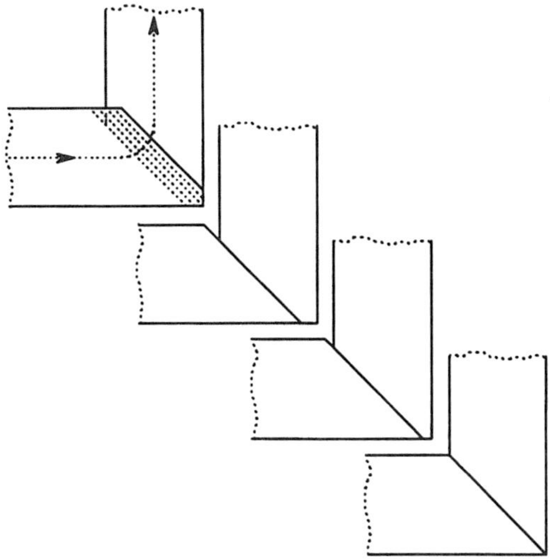

FIGURE 1.13 Successive layers of a laminated core using mitered construction.

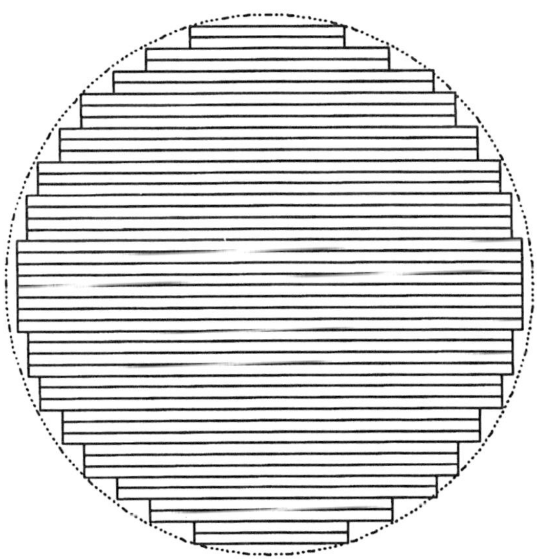

FIGURE 1.14 Development of a circular cross-sectional core.

1.11 MAGNETOSTRICTION

One of the major problems with large power transformers is audible noise that can be loud enough to be very annoying. Essentially all transformer noise is due to a phenomenon called magnetostriction. When a strip of steel is magnetized, it contracts very slightly. At the flux densities used in large power transformers, the amount of magnetostriction is only about 60 μ per meter of length. For a 60 Hz transformer, this small change in dimension occurs 120 times per second. Due to the fact that magnetostriction is not linear with respect to the flux density B there are also harmonics of 120 Hz present in the noise. If any part of the transformer is in resonance with any of the harmonics, the noise can be amplified hundreds of times. Therefore, part of the core design and the overall transformer design is an analysis of the resonant frequencies.

1.12 COMPLETING THE TRANSFORMER BY ADDING A SECOND WINDING

We are now prepared to complete our prototype transformer by adding a second winding. It should now be apparent that by coiling the primary conductor around the core using many turns, a considerable voltage can be induced with only a tiny magnetizing current *if* the peak flux density is kept below the saturation value. Now suppose a *secondary* coil is wound around the same core, surrounding the same magnetic flux as the primary coil as shown in Figure 1.15. Since the secondary coil encircles the same flux as the primary

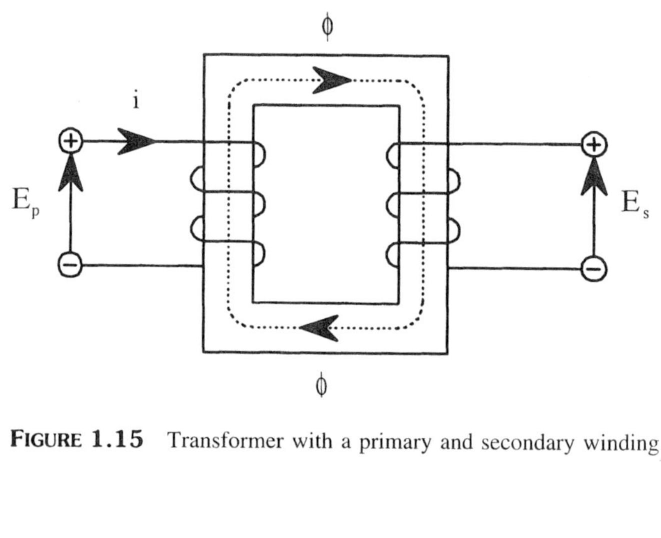

FIGURE 1.15 Transformer with a primary and secondary winding.

coil, the induced voltage per turn is the same in both the primary and secondary coils. Let E_p = applied primary voltage, E_s = induced secondary voltage, N_p = number of primary turns, and N_s = number of secondary turns. Since the induced voltage in the primary coil equals the applied voltage and since the induced volts per turn is the same for both primary and secondary,

$$\frac{E_p}{N_p} = \frac{E_s}{N_s} \tag{1.12.1}$$

$$\frac{E_p}{E_s} = \frac{N_p}{N_s} \tag{1.12.2}$$

Equation (1.12.2) is the first transformer law. The ratio N_p/N_s is called the *transformer turns ratio*, or TTR. Now suppose a load resistance is connected to the secondary coil, as shown in the Figure 1.16. The arrows indicate the directions of the primary and secondary currents. The secondary voltage causes current in the secondary to flow in a direction that always tends to cancel flux in the core. This tendency to cancel flux reduces the induced voltages in both the primary and secondary coils. With a reduction in the induced voltage in the primary, the applied voltage across the primary winding increases the primary current to restore the flux to its original value. Equilibrium is established when the total MMF is just sufficient to induce a voltage equal to the voltage applied across the primary coil. This is equal to the magnetizing MMF:

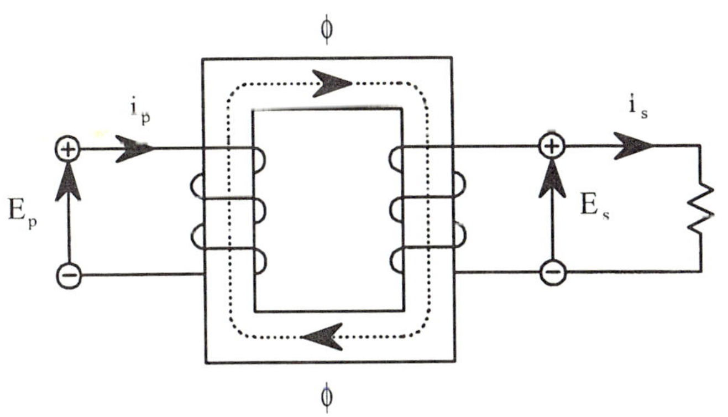

FIGURE 1.16 Two-winding transformer with a load connected to the secondary winding.

$$\text{MMF}_{\text{primary}} + \text{MMF}_{\text{secondary}} = \text{MMF}_{\text{magnetizing}} \tag{1.12.3}$$

With substantial currents flowing in both the primary and secondary winding under load, the magnetizing MMF is negligible compared to the primary MMF and secondary MMF:

$$\text{MMF}_{\text{primary}} + \text{MMF}_{\text{secondary}} = 0 \tag{1.12.4}$$

$$N_p \times i_p + N_s \times i_s = 0 \tag{1.12.5}$$

$$\frac{i_p}{i_s} = \frac{-N_s}{N_p} \tag{1.12.6}$$

The minus sign in Eq. (1.12.6) indicates that the currents in the primary and secondary windings are in opposite directions with respect to producing magnetic flux in the core. While this is undeniably true, the minus sign can be dropped in order to express the *ratio* of primary current to secondary current.

$$\frac{i_p}{i_s} = \frac{N_s}{N_p} \tag{1.12.7}$$

Equation (1.12.7) is the second transformer law. By multiplying E_p/E_s from the first law by $i_{p/}i_s$ from the second law,

$$\frac{E_p \times i_p}{E_s \times i_s} = \frac{N_p \times N_s}{N_s \times N_p} = 1 \tag{1.12.8}$$

$$(E \times i)_p = (E \times i)_s \tag{1.12.9}$$

The meaning of Eq. (1.12.9) is that the instantaneous power supplied to the primary coil from the applied voltage equals the instantaneous power supplied by the secondary coil to the load; in other words, power in = power out. The transformer laws are true irrespective of the phase angles of the currents. Therefore, the real power, reactive power, and apparent power of the primary circuit are all equal to the real power, reactive power, and apparent power of the secondary circuit.

1.13 PLACEMENT OF THE WINDINGS

Schematically, we have depicted the primary and secondary windings as being wound around a common core but located on opposite core legs. In the Chapter 2 we will discuss two very different types of transformers, namely, the core form and the shell form designs. In any transformer design, however, the pri-

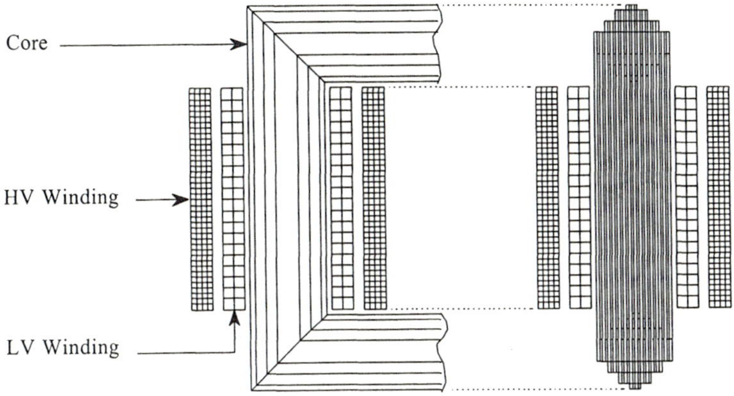

FIGURE 1.17 Cross section of a transformer with primary and secondary windings on a common circular core leg.

mary and secondary windings are always mounted in close proximity to each other in order to maximize the mutual coupling between the windings and thereby increase the overall efficiency. Figure 1.17 illustrates part of a two-winding core form transformer as a cut away view from the side and the end. This configuration has one set of low-voltage and high-voltage windings mounted over a vertical core leg. Note that the core leg and the top and bottom core yokes are stepped to approximate a circular cross section. The laminations are too thin to be seen individually in the edge view. By convention, the HV winding is usually called the primary and the LV winding is called the secondary; however, either the HV or the LV winding can be the input winding.

So far, we have discussed the relationships between voltages and currents for only one pair of primary and secondary windings. In Chapter 2 we will discuss the various. winding connections and transformer configurations with multiple sets of windings used in three-phase and two phase systems.

Example 1.5

A two-winding transformer has a primary winding with 208 turns and a secondary winding with 6 turns. The primary winding is connected to a 4160V system. What is the secondary voltage at no load? What is the current in the primary winding with a 50-amp load connected to the secondary winding? What is the apparent power flowing in the primary and secondary circuits?

First transformer law:

$$\frac{E_p}{E_s} = \frac{N_p}{N_s} = \frac{208}{6}$$

$$E_p = 4160 \text{ V}$$

$$E_s = 4160 \text{ V} \times \frac{6}{208} = 120 \text{ V}$$

Second transformer law:

$$\frac{i_p}{i_s} = \frac{N_s}{N_p} = \frac{6}{208}$$

$$i_s = 50 \text{ A}$$

$$i_p = 50 \text{ A} \times \frac{6}{208} = 1.44 \text{ A}$$

$$\begin{aligned}\text{Apparent power in primary circuit} &= E_p \times I_p \\ &= 4160 \text{ V} \times 1.442 \text{ A} = 5999 \text{ VA}\\ \text{Apparent power in secondary circuit} &= 120 \text{ V} \times 50 \text{ A} = 6000 \text{ VA}\end{aligned}$$

REFERENCES

1. Institute of Electrical and Electronics Engineers, Inc. IEEE Standard Dictionary of Electrical and Electronics Terms, IEEE Std. 100-1972.)
2. R. L. Bean, N. Chackan, Jr., H. R. Moore, and E. C. Wentz. *Transformers for the Electric Power Industry*. McGraw-Hill, New York, 1959, p. 97.

2

Two-Winding Transformer Connections

2.1 INTRODUCTION

Modern electrical systems are almost exclusively three-phase systems, notwithstanding the many miles of distribution circuits that are configured as single-phase taps off of these systems. In addition, there still exist remnants of two-phase systems (typically in mining operations) that were fairly common years ago. When two polyphase systems have different voltages and/or phase angles, these systems can be interconnected using transformers having various possible types of *connections*. Any one of these connections can be accomplished either with a bank of single-phase transformers or by a single polyphase transformer. As we shall see in this chapter, it is in fact possible to interconnect two polyphase systems having a different number of phases using special transformer connections.

A single-phase two-winding transformer is nothing more than a primary and a secondary winding wound around the same magnetic core. Single-phase two-winding transformers can be used in either single-phase circuits or polyphase circuits. A polyphase two-winding transformer contains a number of sets of primary and secondary windings. Each set wound around a separate magnetic core leg. A three-phase two-winding transformer has three sets of primary and secondary windings, and a two-phase two-winding transformer has two sets of primary and secondary windings.

Chapter 1 described the basic theory of operation of a two-winding transformer, and the transformer laws were developed. In this chapter, the principles of the two-winding transformer are applied to polyphase systems. As we shall see, there are a number of possible ways two-winding transformers can be connected to polyphase systems and a number of possible ways that polyphase two-winding transformers can be constructed.

2.2 THE Y-Y CONNECTION IN THREE-PHASE SYSTEMS

The most obvious way of transforming voltages and currents in a three-phase electrical system is to operate each phase as a separate single-phase system. This requires a four-wire system comprised of three phase wires plus a common neutral wire that is shared among the three phases. Each phase is transformed through a set of primary and secondary windings connected phase-to-neutral. This is commonly referred to as the Y-Y connection, as illustrated in Figure 2.1. The left-hand part of Figure 2.1 shows the physical winding connections as three separate two-winding transformers. Both the primary and secondary windings of each of these transformers are connected between one

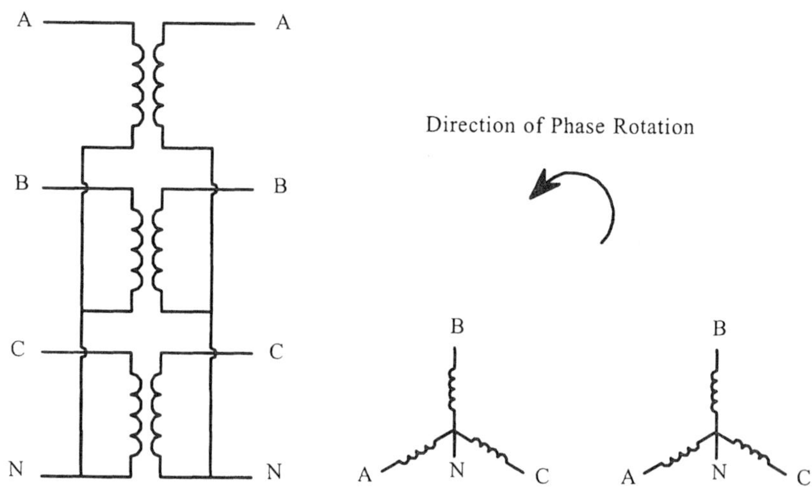

FIGURE 2.1 Y-Y transformer connection and vector diagram.

phase, labeled *A*, *B*, and *C*, and the neutral, labeled *N*. The right-hand part of Figure 2.1 shows the winding connections as a vector diagram. The direction of the phase rotation is assumed to be *A-B-C* expressed in a *counterclockwise direction*. This means that when the vector diagram rotates in a counterclockwise direction on the page, a stationary observer sees *A* phase, followed by *B* phase, and followed by *C* phase in sequence. This counterclockwise convention will be followed throughout this book.

The term "Y-Y connection" should be obvious from the fact that the vector diagrams of the primary and secondary windings both resemble the letter Y. Each phase of the primary and secondary circuits is 120 electrical degrees out of phase with the other two phases. This is represented by angles of 120° between the legs of the primary Y and the secondary Y in the vector diagram. Each primary winding is magnetically linked to one secondary winding through a common core leg. Sets of windings that are magnetically linked are drawn parallel to each other in the vector diagram. In the Y-Y connection, each primary and secondary winding is connected to a neutral point. The neutral point may or may not be brought out to an external physical connection and the neutral may or may not be grounded.

As we saw in Chapter 1, transformer magnetizing currents are not purely sinusoidal, even if the exciting voltages are sinusoidal. The magnetizing currents have significant quantities of odd-harmonic components. If three identical transformers are connected to each phase and are excited by 60 Hz voltages of equal magnitude, the 60 Hz fundamental components of the exciting currents cancel out each other at the neutral. This is because the 60 Hz fundamental currents of *A*, *B*, and *C* phase are 120° out of phase with one another and the vector sum of these currents is zero. The third, ninth, fifteenth and other so-called *zero-sequence* harmonic currents are *in phase* with each other; therefore, these components do not cancel out each other at the neutral but add in phase with one another to produce a zero-sequence neutral current, provided there is a path for the neutral current to flow.

Due to the nonlinear shape of the *B-H* curve, odd-harmonic magnetizing currents are required to support sinusoidal induced voltages. If some of the magnetizing current harmonics are not present, then the induced voltages cannot be sinusoidal. If the neutrals of both the primary and the secondary are open-circuited and there is no path for the zero-sequence harmonic currents to flow, the induced voltages will not be sinusoidal.

Figure 2.2 depicts the situation where the primary neutral is returned to the voltage source in a four-wire three-phase circuit. Each of the magnetizing currents labeled i_A, i_B, and i_C contain the 60 Hz fundamental current and all of the odd harmonic currents necessary to support sinusoidal induced voltages.

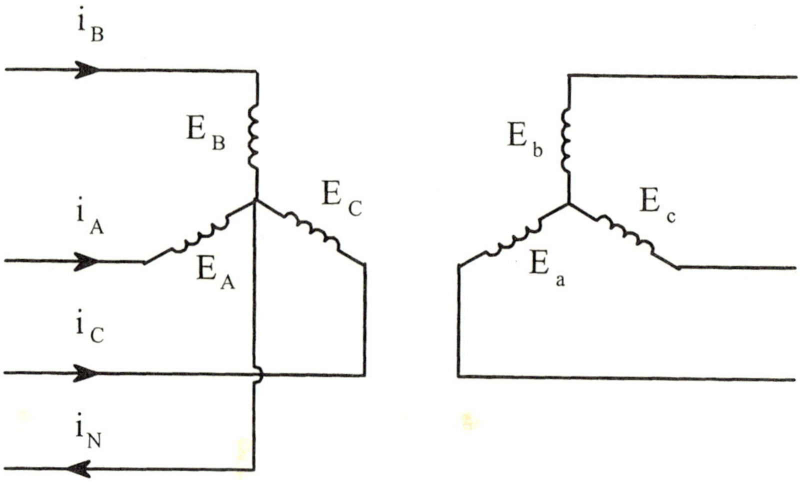

FIGURE 2.2 Y-Y Connection with the primary neutral brought out.

The zero-sequence magnetizing currents combine to form the neutral current i_N, which returns these odd harmonics to the voltage source. Assuming that the primary voltage is sinusoidal, the induced voltages E_A, E_B, and E_C (in both the primary and secondary) are sinusoidal as well.

This situation changes dramatically if the neutrals of both sets of the primary and secondary windings are open-circuited, as shown in Figure 2.3.

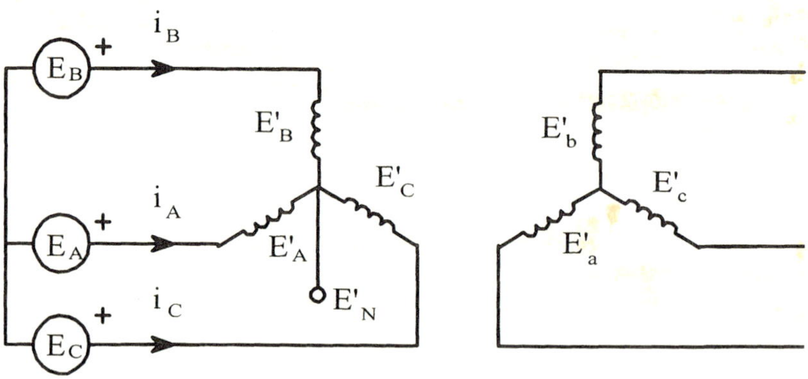

FIGURE 2.3 Voltage at the primary neutral of a Y-Y connection with the primary and secondary neutrals isolated.

Even if the source voltages E_A, E_B, and E_C are purely sinusoidal, the induced voltages E'_A, E'_B, and E'_C will not be sinusoidal. This results in distortions of the secondary voltages. The resulting voltage distortion is equivalent to a Y-Y transformer with zero-sequence currents allowed to flow in the primary neutral with an imaginary superimposed primary winding carrying only the zero-sequence currents 180° out of phase with the normal zero-sequence currents (see Ref. 1). Analysis of the voltages induced by the "phantom primary windings" is greatly complicated by the fact that the core is highly nonlinear so that each of the individual zero-sequence harmonics currents carried by the phantom primary windings will induce even higher-order harmonic voltages as well.

Fourier analysis can be used to arrive at an approximation of the secondary voltages with an open primary neutral. Taking one phase at a time, the normal magnetizing current for a sinusoidal exciting voltage is plotted from the *B-H* curve of the transformer using the technique outlined in Chapter 1. The normal magnetizing current is converted to a Fourier series and then it is reconstructed by removing all of the zero-sequence harmonics. The resulting exciting current will have a shape different from the normal exciting current, which is then used to construct an induced voltage using the *B-H* curve in the reverse manner that was used to construct the original exciting current. This process is rather laborious, so suffice it to say that if a Y-Y transformer does not have a neutral path for zero-sequence exciting currents, there will be harmonic voltages induced in the secondary even if the exciting voltage is purely sinusoidal.

Another problem with a Y-Y connection is that if an unbalanced phase-to-neutral load is connected to the secondary, the phase voltages at the load will be unbalanced also unless the neutral of the load is connected to the secondary neutral of the transformer. Therefore, it is generally a good practice to bring out and ground both the primary and the secondary neutrals in a Y-Y transformer connection.

2.3 ADVANTAGES OF THE Y-Y CONNECTION

Although care must be exercised when using the Y-Y connection, this connection has certain inherent and important advantages over other three-phase transformer connections.

1. The primary and secondary circuits are *in phase*; i.e., there are no phase angle displacements introduced by the Y-Y connection. This is an important advantage when transformers are used to interconnect systems of different voltages in a cascading manner. For exam-

ple, suppose there are four systems operating at 500, 230, 138, and 69 kV that need to be interconnected. Substations can be constructed using Y-Y transformer connections to interconnect any two of these voltages. The 500 kV system can be tied with the 69 kV system through a single 500 to 69 kV transformation or through a series of cascading transformations at 230, 138, and 69 kV.

2. Since the phase-to-neutral voltage is only 57.7% of the phase-to-phase voltage, the windings of a Y-Y transformer require fewer turns to produce the same level of excitation in the core compared to windings connected across the phases.
3. If the neutral end of a Y-connected winding is grounded, then there is an opportunity to use reduced levels of insulation at the neutral end of the winding. A winding that is connected across the phases requires full insulation throughout the winding.
4. A Y-Y transformer may be constructed as an *autotransformer*, with the possibility of great cost savings compared to the two-winding transformer construction. (Chapter 4 will fully develop the topic of autotransformers.)

2.4 DISADVANTAGES OF THE Y-Y CONNECTION

The Y-Y transformer connection was poorly understood in the early days of power engineering and it received a very bad reputation when it was first used; in fact, this connection was avoided for a long time until its limitations were overcome by good engineering practice. Some of the inherent disadvantages of the Y-Y connection are discussed below:

1. The presence of third (and other zero-sequence) harmonics at an ungrounded neutral can cause overvoltage conditions at light load. When constructing a Y-Y transformer using single-phase transformers connected in a bank, the measured line-to-neutral voltages are not 57.7% of the system phase-to-phase voltage at no load but are about 68% and diminish very rapidly as the bank is loaded. The effective values of voltages at different frequencies combine by taking the square root of the sum of the voltages squared. With sinusoidal phase-to-phase voltage, the third-harmonic component of the phase-to-neutral voltage is about 60%, so the effective voltage across the winding is calculated as follows:

   $E = [0.577^2 + (0.6 \times 0.577)^2]^{1/2} = 68\%$

2. There can be a large voltage drop for unbalanced phase-to-neutral loads. This is caused by the fact that phase-to-phase loads cause a voltage drop through the *leakage reactance* of the transformer whereas phase-to-neutral loads cause a voltage drop through the *magnetizing reactance*, which is 100 to 1000 times larger than the leakage reactance. (The subject of leakage reactance will be discussed in Chapter 3.)
3. Under certain circumstances, a Y-Y connected three-phase trans-

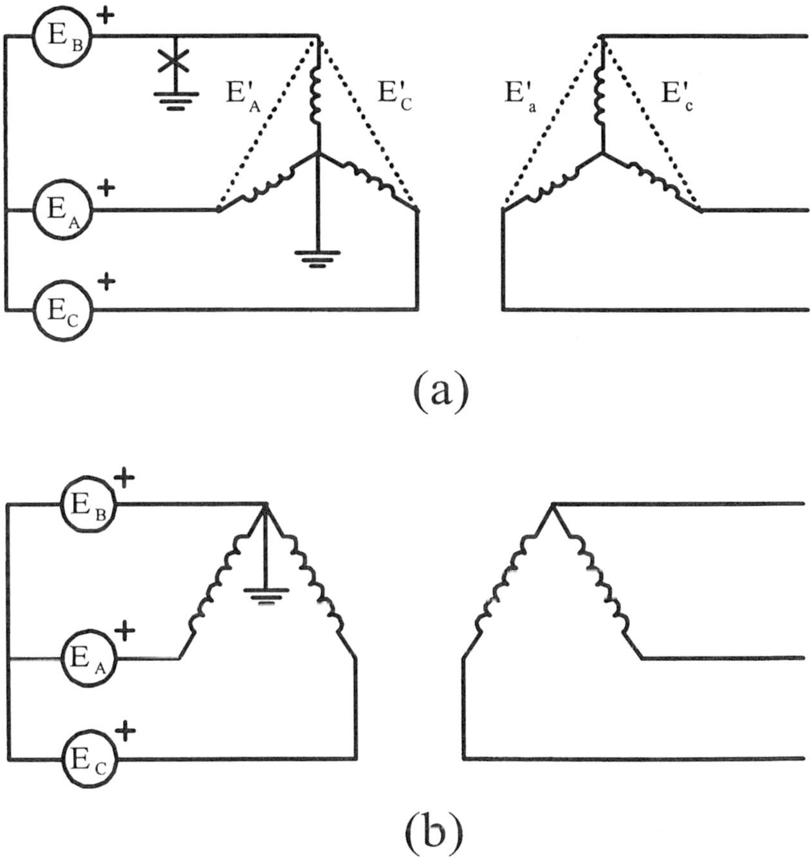

FIGURE 2.4 Overvoltage condition produced by a fault at the primary of a Y-Y connection.

former can produce severe tank overheating that can quickly destroy the transformer. This usually occurs with an open phase on the primary circuit and load on the secondary. (The tank heating phenomenon will be discussed at length in Chapter 5.)

4. Series resonance between the third harmonic magnetizing reactance of the transformer and line-to-ground capacitance can result in severe overvoltages. (This phenomenon, called *ferroresonance*, will be discussed in Chapter 5.)
5. If a phase-to-ground fault occurs on the primary circuit with the primary neutral grounded, then the phase-to-neutral voltage on the unfaulted phases increases to 173% of the normal voltage. This would almost certainly result in overexcitation of the core, with greatly increased magnetizing currents and core losses. This is illustrated in Figure 2.4. A bold X marks the location of a *B* phase-to-ground fault with the neutral of the voltage source either ungrounded or connected to ground through a large impedance. The voltage across the *B* phase winding collapses and the applied voltages across the *A* phase and *C* phase windings are now equal in magnitude to the phase-to-phase voltages.
6. If the neutrals of the primary and secondary are both brought out, then a phase-to-ground fault on the secondary circuit causes neutral fault current to flow in the primary circuit. Ground protection re-

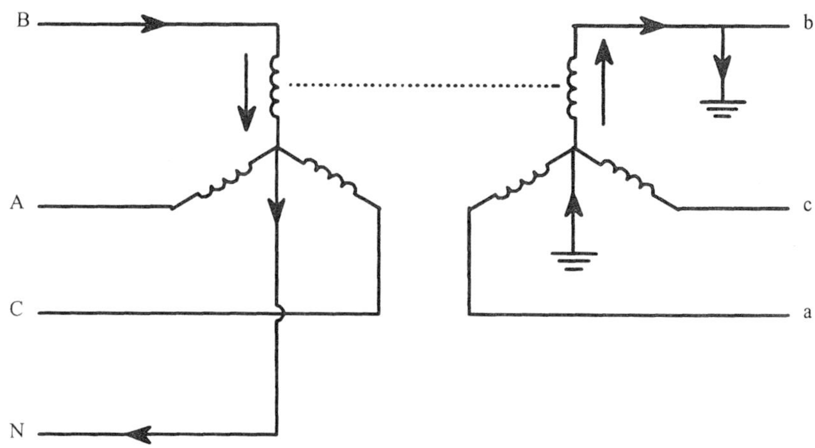

FIGURE 2.5 Primary neutral current for a fault on the secondary side of a Y-Y connection.

laying in the neutral of the primary circuit may then operate for faults on the secondary circuit. This is illustrated in the Figure 2.5. The two magnetically coupled windings are connected by a dotted line. Any fault current in the secondary neutral is transformed into neutral current in the primary circuit through the second transformer law.

The obvious remedy for some of the disadvantages of the Y-Y transformer connection would be to simply solidly ground both the primary and secondary neutrals. In fact, this is standard practice for virtually all Y-Y transformers in systems designed by utility companies. Unfortunately, solidly grounding the neutrals alone does not solve the problem of tank overheating, ferroresonance, and operating primary ground protection during secondary faults.

2.5 THE Y-Δ CONNECTION AND THE Δ-Y CONNECTION

Connecting the primary or secondary winding across the phases of a three-phase transformer can eliminate many of the major drawbacks of the Y-Y connection. A winding connected phase-to-phase is called a *delta-connected winding* because of its resemblance to the Greek letter Δ when it is depicted in a vector diagram. Since a Δ winding has no connection to its neutral point, it is usually left ungrounded, so it ''floats'' above ground potential. The Δ-Y connection is shown in Figure 2.6.

The left-hand part of Figure 2.6 shows the physical winding connections as a set of two-winding transformers. The primary winding of each transformer is connected phase-to-phase and the secondary of each transformer is connected phase-to-neutral with the neutrals grounded. The right-hand part of Figure 2.6 shows the winding connections shown as a vector diagram. As usual, the direction of phase rotation is assumed to be *A-B-C* in a counterclockwise direction.

One of the major advantages of the Δ-Y connection is that it provides *harmonic suppression*. Recall that the magnetizing current must contain odd harmonics for the induced voltages to be sinusoidal and the third harmonic is the dominant harmonic component. In a three-phase system the third harmonic currents of all three phases are in phase with each other because they are zero-sequence currents. In the Y-Y connection, the only path for third harmonic current is through the neutral. In the Δ-Y connection, however, the third harmonic currents, being equal in amplitude and in phase with each other, are

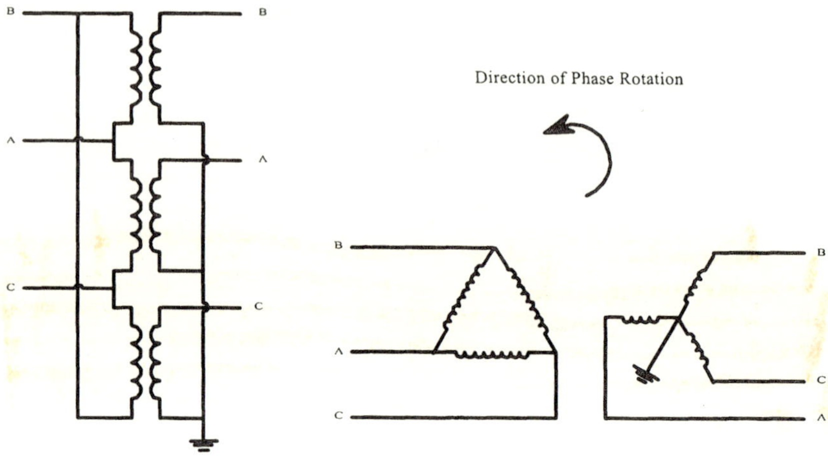

FIGURE 2.6 Δ-Y transformer connection and vector diagram.

able to circulate around the path formed by the Δ-connected winding. The same thing is true for the other zero-sequence harmonics.

If the magnetizing current waveforms of all three phases have approximately the same shape and magnitude, and if the phase angles of the 60 Hz fundamental frequencies are 120° apart, then the third harmonic currents circulate only within the transformer bank and do not emerge as physical currents at the terminals of the windings. This causes these currents to be *suppressed* in the primary and secondary circuits. Typical transformer magnetizing currents for the three different phases are plotted in Figure 2.7. The magnetizing

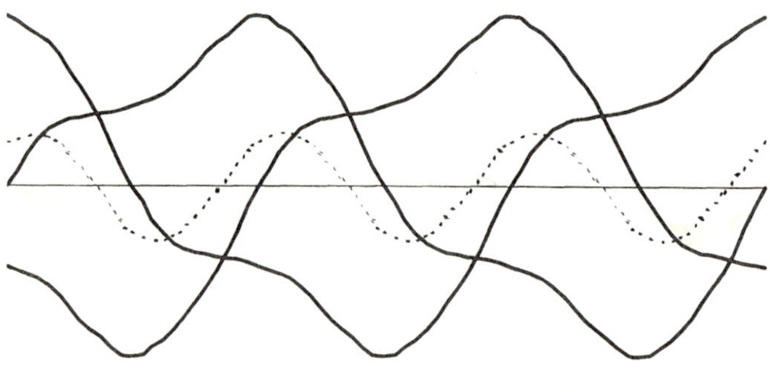

FIGURE 2.7 Three-phase excitation current wave forms (solid) and the third harmonic components combined (dotted).

currents of the three phases are plotted by the heavy lines. For excitation at 60 Hz, the three magnetizing currents are separated in time by $^1/_{180}$ of a second. The phase angles of the 60 Hz fundamental frequencies are electrically by 120°. In the Δ winding, the third harmonic currents of the three waveforms in phase with each other and add algebraically to form the wave form shown in Figure 2.7 as the dotted line with the sinusoidal shape. The combined third harmonic current is plotted true to scale.

Another important advantage of the Δ-Y connection is that it provides *ground current isolation* between the primary and secondary circuits. Assuming that the neutral of the Y-connected secondary circuit is grounded, a load connected phase-to-neutral or a phase-to-ground fault produces two equal and opposite currents in two phases in the primary circuit without any neutral ground current in the primary circuit. Therefore, in contrast with the Y-Y connection, phase-to-ground faults or current unbalance in the secondary circuit will not affect ground protective relaying applied to the primary circuit. This feature enables proper coordination of protective devices and is a very important design consideration.

A transformer-connected Δ-Y with the neutral of the Y grounded is sometimes referred to as a *grounding bank*, because it provides a *local* source of ground current at the secondary that is isolated from the primary circuit. For example, suppose an ungrounded generator supplies a simple radial system through a Δ-Grounded Y transformer as shown Figure 2.8. The generator can supply a single-phase-to-neutral load through the Δ-grounded Y transformer. Let us refer to the low-voltage generator side of the transformer as the secondary and the high-voltage load side of the transformer as the primary. Note that each primary winding is magnetically coupled to a secondary winding. The magnetically coupled windings are drawn in parallel to each other in Figure

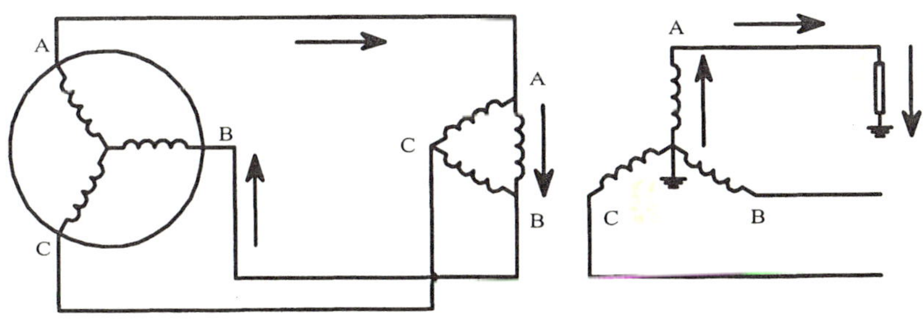

FIGURE 2.8 Ungrounded generator connected to a single-phase load through a Δ-Y transformer.

2.8. Through the second transformer law, the phase-to-ground load current in the primary circuit is reflected as a current in the *A*-*C* secondary winding. No other currents are required to flow in the *A*-*C* or *B*-*C* windings on the generator side of the transformer in order to balance ampere-turns.

The Δ-Y transformer connection is used universally for connecting generators to transmission systems because of two very important reasons. First of all, generators are usually equipped with sensitive ground fault relay protection. The Δ-Y transformer is a source of ground currents for loads and faults on the transmission system, yet the generator ground fault protection is completely isolated from ground currents on the primary side of the transformer. Second, rotating machines can literally be shaken apart by mechanical forces resulting from zero-sequence currents. The Δ-connected winding blocks zero-sequence currents on the transmission system from the generator.

2.6 PHASE ANGLE DISPLACEMENT AND PHASE ROTATION

A careful examination of the primary and secondary voltage vectors in Figure 2.6 reveals that in a Δ-Y connection the primary voltage of any given phase is displaced electrically by 30° from the secondary voltage of the corresponding phase. The 30° phase angle displacement may either be positive or negative; i.e., the primary voltage may either *lead* the secondary voltage by 30° or *lag* the secondary voltage by 30° depending on how the transformer is connected. The standard connection for a Δ-Y transformer is to have the high-side voltages *lead* the low-side voltages by 30°, as shown in Figure 2.9. A dotted line is shown connecting the *A* phase connection of the Δ to an imaginary neutral point within the Δ. This neutral point is not a physical connection. With the usual *A*-*B*-*C* counterclockwise phase rotation, the primary voltage, labeled *A*, leads the secondary voltage, labeled *a*, by an angle of 30° as shown by the small vector diagram at the bottom of the figure.

The property of a Δ-Y transformer that causes the primary and secondary voltages to be out of phase with each other is referred to as the *phase angle displacement* of the transformer. Phase angle displacement must not be confused with *phase rotation*. Phase rotation is the order in which the phase voltages reach their peak values. In three-phase systems, the phase rotation can be either *A*-*B*-*C* or *C*-*B*-*A*. The phase rotation can be reversed by interchanging any *two* phases of the primary circuit *or* any *two* phases of the secondary circuit. Interchanging two phases on both the primary circuit *and* the secondary circuit will revert back to the original phase rotation.

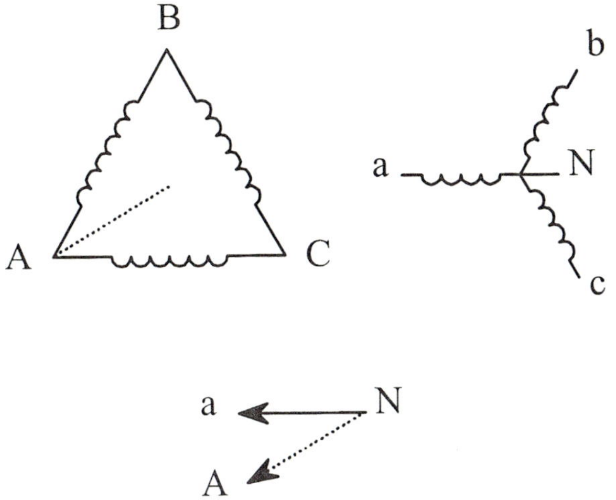

FIGURE 2.9 Phase angle displacement in a Δ-Y transformer.

Interchanging the phase connections in a Δ-Y transformer not only reverses the phase rotation but changes the phase angle displacement from a standard $-30°$ displacement to a $+30°$ displacement, as illustrated in Figure 2.10. The upper part of the figure shows a Δ-Y transformer with the primary terminals labeled 1, 2, and 3 and the secondary terminals labeled 1′, 2′, and 3′. Suppose *A* phase is connected to terminal 1, *B* phase is connected to terminal 2, and *C* phase is connected to terminal 3. Since the transformer has a standard phase displacement, the primary leads the secondary by 30° for an *A-B-C*, 1–2-3 phase rotation. Suppose *A* phase and *C* phase are interchanged on both the primary *and* the secondary sides of the transformer, which retains the original *A B C* phase rotation at the secondary. The transformer is not physically altered; however, in order to produce an *A-B-C*, 3-2-1 phase rotation and retain the convention of a counterclockwise vector rotation used throughout this book, the vector diagram of the transformer must be redrawn as the mirror image of the vector diagram of the transformer with a 1-2-3 phase rotation.

This mirror image is shown in the lower part of Figure 2.10. Note that the relationships between the three sets of windings has not changed electrically or magnetically in the mirror image. In other words, the 1-3 winding is still magnetically linked to the 1′-*N* winding, the 2-1 winding is still magnetically linked to the 2′-*N* winding, and the 3-2 winding is still magnetically

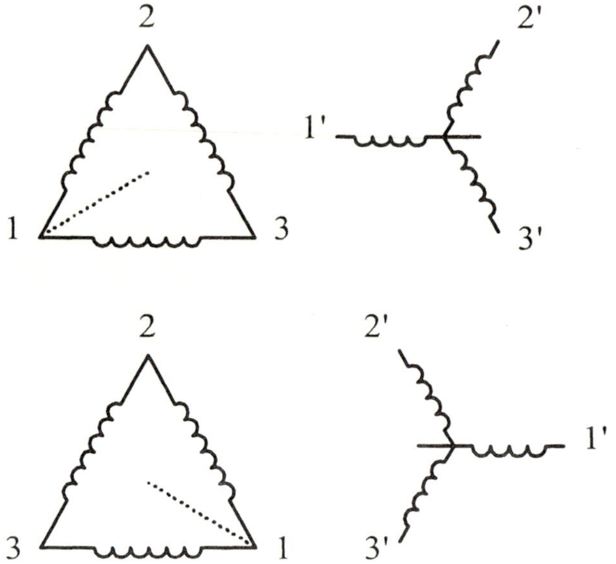

FIGURE 2.10 Phase angle displacements of a Δ-Y transformer for a 1-2-3 phase rotation and a 3-2-1 phase rotation.

linked to the 3′-*N* winding in exactly the same way. What has changed, however, is the phase angle displacement between the primary and the secondary circuits. The primary circuit now *lags* the secondary circuit by 30° with the conventional counterclockwise rotation. Careful attention to phase angle displacement is important; otherwise potentially disastrous results can occur. This is illustrated by the following example, which was based on an actual case.

Example 2.1

A certain major utility company, named TransCo, supplied power to a small municipal distribution company, named DisCo. The DisCo substation has standard Δ-Y connected transformers. Both TransCo and DisCo had an *A-B-C* phase rotation. The voltages on the DisCo distribution system lagged TransCo transmission voltages by 30° due to the phase displacement of the Δ-Y transformers.

TransCo is directly interconnected with another major utility company, named WireCo. The phase rotation of WireCo is *C-B-A* throughout its transmission and distribution systems. Nevertheless, the TransCo and WireCo transmission systems are interconnected at various places, simply by swapping the phases labeled *A* and *C* at the point of interconnection, reversing the phase rotations at the points of interconnection. WireCo supplies its own distribution system through substations that have standard Δ-Y transformers. Because of the nonstandard *C-B-A*, 3-2-1 phase rotation, however, WireCo's distribution system voltages *lead* its transmission system voltages by 30°.

Many years go by and WireCo purchases the small municipal distribution company, DisCo. WireCo plans to run a transmission line to supply DisCo's substation, and disconnect the transmission line from TransCo. Because DisCo has an *A-B-C* phase rotation, whereas all of WireCo has a *C-B-A* phase rotation, it was recognized that some kind of a phase swap must be made in order to maintain the *A-B-C* phase for DisCo's distribution customers. There was also a future plan to integrate the DisCo distribution system with the WireCo distribution system by building distribution ties between the two systems. The fact that DisCo had an *A-B-C* phase rotation while WireCo had a *C-B-A* phase rotation was not seen as a problem, since phase rotations can be reversed simply by swapping two phases.

The big day arrived when DisCo's transmission supply was converted from the original supplier TransCo to the new supplier WireCo. Figure 2.11 shows the configuration that WireCo installed to accomplish the conversion. The *A* and *C* phases of WireCo's transmission line were swapped at the high-voltage side of the Δ-Y transformers in DisCo's substation. This provided the proper *A-B-C* phase rotation for DisCo's customers, but it was later realized that no amount of phase swapping would accomplish interconnecting DisCo's

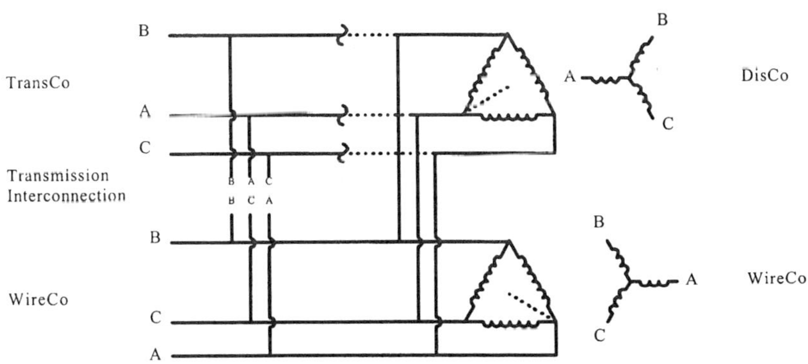

FIGURE 2.11 Improper connections resulting in a 60° phase displacement between distribution systems.

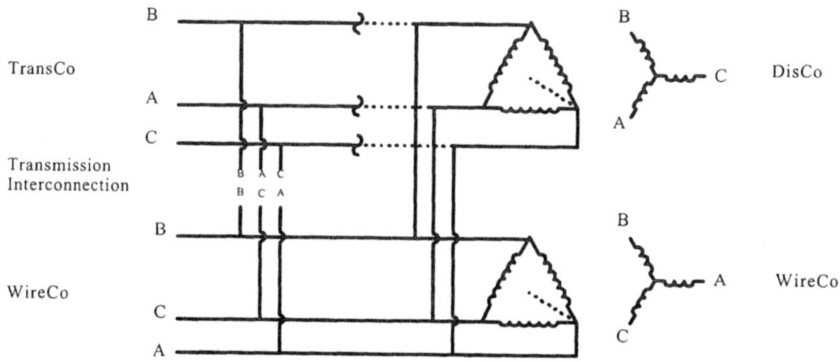

FIGURE 2.12 Proper connections resulting in 0° phase displacement between distribution systems.

distribution system with WireCo's distribution system, making it impossible to fully integrate the two systems. What had gone wrong?

It is clear by studying the vector relationships in Figure 2.11 that there is a permanent 60° phase angle displacement between the DisCo and WireCo distribution systems. This happened because both standard and nonstandard transformer connections were used at the substations. In order to properly tie together the two systems, WireCo should have connected its transmission lines to the transformers at DisCo's substation in the same way as the connections are made at WireCo's substations, as depicted in Figure 2.12.

The connections shown in Figure 2.12 would have resulted in 3–2-1 rotations for both sets of transformers, and also the same phase displacements. To provide the required *A-B-C* phase rotation for DisCo's distribution system, *A* and *C* phases could be swapped at the *low* side connections to the DisCo substation transformers. The resulting voltages would now be in-phase with WireCo's distribution system, albeit with a reverse phase rotation. Now all that is required to interconnect the DisCo and WireCo distribution systems is to swap the phases labeled *A* and *C* at the point of interconnection.

2.7 THE Y-Δ GROUNDING BANK

There are times when a grounded Y-Δ transformer is used for no other purpose than to provide a good ground source in an otherwise ungrounded system. Take, for example, a distribution system supplied by a Δ-connected (i.e., ungrounded) power source. If it is required to connect phase-to-ground loads to

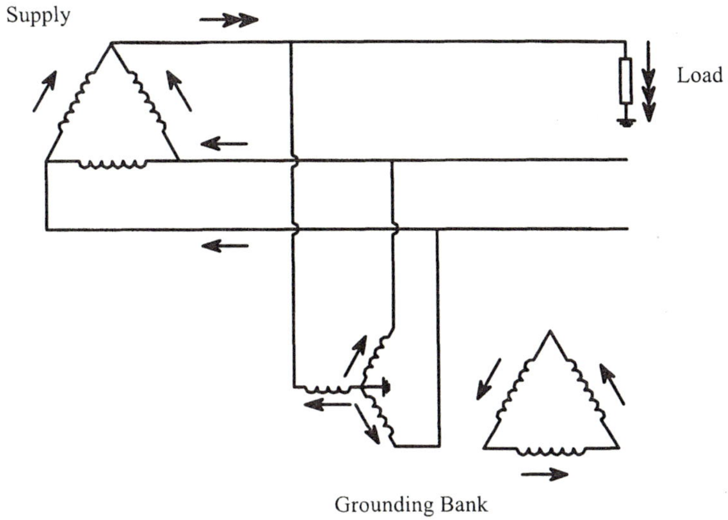

FIGURE 2.13 Δ-Y transformer as a local source of ground current (grounding bank).

this system a grounding bank is connected to the system, as shown in Figure 2.13. Note that the Δ-connected winding is not connected to any external circuit in Figure 2.13. With a load current equal to 3 times i, each phase of the grounded Y winding provides the same current i, with the Δ-connected secondary winding of the grounding bank providing the ampere-turns required to cancel the ampere-turns of the primary winding. Note that the grounding bank does not supply any real power to the load; it is there merely to provide a ground path. All the power required by the load is supplied by two phases of the ungrounded supply.

Example 2.2

A 13.8 kV Δ-connected power supply is connected to a 30-A single phase-to-ground load with a grounding bank. How much KVA does the Δ-connected power supply provide?

From the diagram in Figure 2.13, two of the power supply windings carry one-third of the phase-to-ground current each, or 10 A each. Since the phase-to-phase voltage is 13.8 kV, each of these windings supplies 13.8 kV × 10 A = 138 KVA. The third winding carries no current. Thus, the power supply provides 2 × 138 KVA = 276 KVA.

The phase-to-ground voltage is 13.8 kV/1.732 = 7.967 kV. Thus, the KVA consumed by the 30 A load is 7.967 kV × 30 A = 239 KVA. How do we explain the apparent "excess" KVA from the power supply? Suppose the load is connected between *A* phase and ground and let us assume the load current is in phase with its phase-to-neutral voltage, both having a 0° angle. Therefore, the power consumed by the load is 239 kW.

The load current is comprised of $10\angle 0°$ A from the *A-B* winding, $10\angle 0°$ A from the *A-C* winding, and $10\angle 0°$ A from *A* phase of the grounding bank. The *A*-to-*B* phase voltage is $13.8\angle 30°$ kV and the *A*-to-*C* phase voltage is $13.8\angle -30°$ kV. The power outputs from the two windings are calculated below.

$$S_{ab} = E_{ab} \times i^*_{ab}$$

$$S_{ac} = E_{ac} \times i^*_{ac}$$

where i^*_{ab} is the complex conjugate of the current i_{ab} and i^*_{ac} is the complex conjugate of the current i_{ac}.

$$S_{ab} = 13{,}800\angle 30° = 10\angle 0° = 119.5 \text{ kW} + 69 \text{ KVAR}$$

$$S_{ab} = 13{,}800\angle -30° = 10\angle 0° = 119.5 \text{ kW} - 69 \text{ KVAR}$$

The total real power output is the same as the real power consumed, or 239 kW. The "excess" KVA output from the power supply is due to 69 KVAR of reactive power that circulates within the circuit.

2.8 THE ZIGZAG CONNECTION

The zigzag connection is also called the *interconnected star connection*. This connection has some of the features of the Y and the Δ connections, combining the advantages of both. The zigzag connection is a three-phase connection and is constructed as shown in Figure 2.14. There are three pairs of windings, each having a 1:1 turns ratio. The left-hand set of windings shown in the figure is a conventional Y connection, *a′-b′-c′*, with the neutral *N* brought out. The open ends of the Y are electrically connected to the right-hand set of windings as follows: *a′* connects to the right-hand winding paired with to the *b′-N* winding, *b′* connects to the right-hand winding paired to *c′-N* winding, and *c′* connects to the right-hand winding paired to the *a′-N* winding. The opposite ends of the right-hand windings are brought out as the phase terminals *a*, *b*, and *c*.

The vector diagram shown on the right of Figure 2.14 makes it is obvious why this is called a zigzag connection. It operates on the following principle:

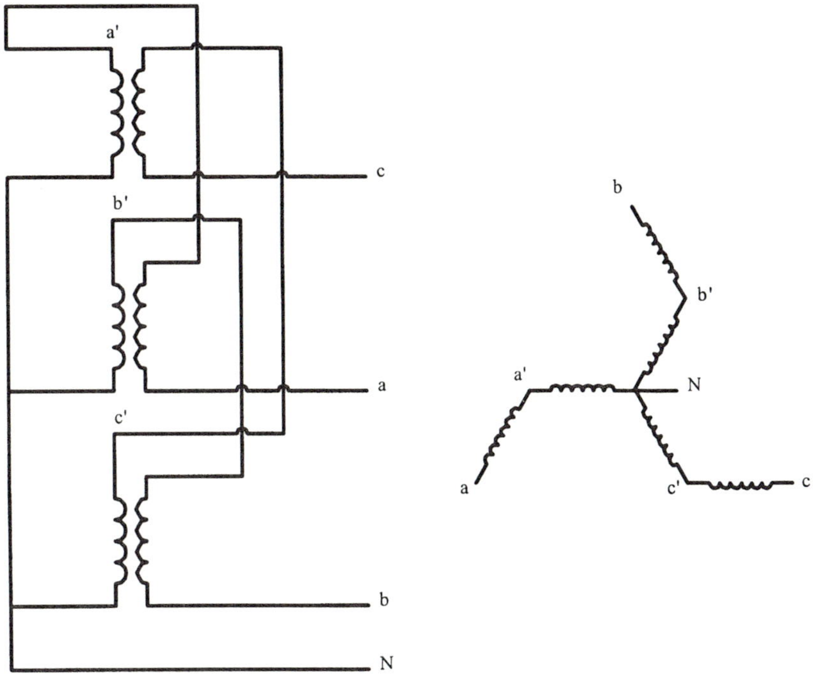

FIGURE 2.14 Interconnected star or zigzag winding connection and vector diagram.

If three currents, equal in magnitude and phase, are applied to the three terminals, the ampere-turns of the a'-N winding cancel the ampere-turns of the c'-c winding, the ampere-turns of the b'-N winding cancel the ampere turns of the a'-a winding, and the ampere-turns of the c'-N winding cancel the ampere turns of the b'-b winding. Therefore, the transformer allows the three in-phase currents to easily flow to neutral.

If three currents, equal in magnitude but 120° out of phase with each other, are applied to the three terminals, the ampere-turns in the windings cannot cancel and the transformer restricts the current flow to the negligible level of magnetizing current. Therefore, the zigzag winding provides an easy path for in-phase currents but does not allow the flow of currents that are 120° out of phase with each other.

The ability to provide a path for in-phase currents enables us to use the zigzag connection as a grounding bank, which is one of the main applications for this connection. If a zigzag winding is used as a secondary winding with

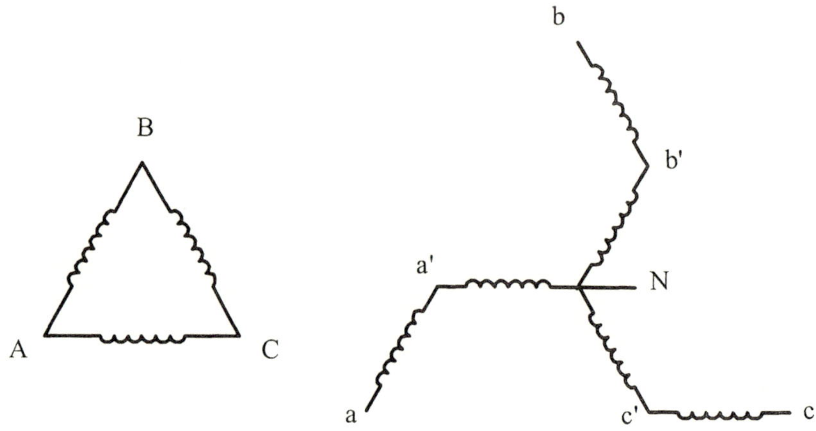

FIGURE 2.15 Δ-zigzag transformer vector diagram.

a Δ winding used as a primary winding, the Δ-zigzag connection is created, as shown in Figure 2.15. A Δ-zigzag transformer is technically not a two-winding transformer but rather a three-winding transformer because three separate windings are wound around each core leg. Three winding transformers will be discussed in greater detail in Chapter 4. Since two of the sets of windings are interconnected, we treat the Δ-zigzag as if it were a two-winding transformer. As usual, the sets of windings that are magnetically linked on common core legs are drawn in parallel to each other, as shown in Figure 2.15.

The Δ-zigzag connection provides the same advantages as the Δ-Y connection, like third harmonic suppression and ground current isolation. One added advantage is that there is no phase angle displacement between the primary and the secondary circuits with this connection; therefore, the Δ-zigzag connection can be used in the same manner as Y-Y and Δ-Δ transformers without introducing any phase shifts in the circuits.

2.9 COMPARISONS OF ECONOMY OF THE DIFFERENT WINDING CONFIGURATIONS

Some of the advantages and disadvantages of the Y, Δ, and zigzag connections have been discussed in terms of voltage distortion, ground current isolation, etc., but their performance under load still needs to be addressed. A transformer's load-carrying capability is determined primarily by the total I^2R losses in the windings, so an obvious way to reduce losses is to reduce the winding currents. Since the KVA transformed is equal to the voltage across the winding times the winding current, increasing the voltage greatly decreases I^2R losses for a given KVA transformed.

The resistance of the winding is proportional to its conductor length and is inversely proportional to the conductor cross section area. Both resistance and the losses are reduced if the total conductor length is reduced. For a given set of core dimensions, however, the number of turns must be large enough to limit the volts per turn to levels below saturation; therefore, the length of the conductor must also be proportional to the voltage across the winding.

To compare the quantity of copper conductor required in transformers using Y, Δ, and zigzag windings, assume that all three transformers operate at the same phase-to-phase voltages, all three transform the same KVA, and that all three core configurations are the same. The objective is to keep the I^2R losses constant by varying the amounts of materials used.

The phase-to-phase voltage across the Δ winding is 1.732 times the phase-to-ground voltage across the Y winding; therefore, the Δ winding conductor must be 1.732 times longer than an equivalent Y winding conductor to limit the volts per turn. The load current through the Δ winding is 0.577 times the load current through the Y winding for the same KVA. The losses for each winding are calculated as follows.

$$(I^2R)_Y = (1.0 \times i_Y)^2 \times \rho \times 1.0 \times \frac{L_Y}{A_Y} \tag{2.9.1}$$

$$(I^2R)_\Delta = (0.577 \times i_Y)^2 \times \rho \times 1.732 \times \frac{L_Y}{A_\Delta} \tag{2.9.2}$$

where

i_Y = current in the Y winding
ρ = resistivity of copper
L_Y = conductor length of the Y winding
A_Y = cross-sectional area of the Y winding conductor
A_Δ = cross-sectional area of the Δ winding conductor

Since the I^2R losses are to be the same for both windings, $(I^2R)_Y = (I^2R)_\Delta$ and

$$\frac{1}{A_Y} = \frac{0.577}{A_\Delta} \tag{2.9.3}$$

$$A_\Delta = 0.577 \times A_Y \tag{2.9.4}$$

The total volume of the copper conductor is equal to its length times its cross-section area:

$$V_Y = L_Y \times A_Y \tag{2.9.5}$$

$$V_\Delta = 1.732 \times L_Y \times 0.577 \times A_Y = V_Y \tag{2.9.6}$$

Therefore, the quantity of copper used in a Δ winding is about the same as the quantity of copper used in an equivalent Y winding.

The zigzag winding is comprised of two windings in series. These series windings are on different core legs and have different induced voltages. But since the current through the two windings in series is equal to the phase-to-neutral current, the current through the zigzag winding is the same as the current through the Y winding. The voltage across each series winding in the zigzag connection is 0.577 times the phase-to-neutral voltage; therefore, the total length of conductor per phase in the zigzag winding is $2 \times 0.577 = 1.155$ times the conductor length of the equivalent Y winding. The losses for each winding are calculated as follows.

$$(I^2R)_Y = (1.0 \times i_Y)^2 \times \rho \times 1.0 \times \frac{L_Y}{A_Y} \tag{2.9.7}$$

$$(I^2R)_{ZZ} = (1.0 \times i_Y)^2 \times \rho \times 1.155 \times \frac{L_Y}{A_{ZZ}} \tag{2.9.8}$$

where

i_Y = Y winding current
ρ = resistivity of copper
L_Y = conductor length of the Y winding
A_Y = cross-sectional area of the Y winding conductor
A_{ZZ} = cross-sectional area of the zigzag winding conductor

Since the I^2R losses are to be the same for both windings, $(I^2R)_Y = (I^2R)_{ZZ}$ and

$$\frac{1}{A_Y} = \frac{1.155}{A_{ZZ}} \tag{2.9.9}$$

$$A_{ZZ} = 1.155 \times A_Y \tag{2.9.10}$$

The total volume of the copper conductor is equal to its length times its cross-sectional area:

$$V_Y = L_Y \times A_Y \tag{2.9.11}$$

$$V_{ZZ} = 1.155 \times L_Y \times 1.155 \times A_Y = 1.33 \times V_Y \tag{2.9.12}$$

Therefore, the quantity of copper required in a zigzag winding is about 33% greater than the quantity of copper used in an equivalent Y winding having the same KVA rating. The added cost of copper in a zigzag winding and the increased complexity of its design limit the use of zigzag windings to very special applications where the unique advantages of the electrical characteristics outweigh the disadvantages.

2.10 TRADE-OFF BETWEEN STEEL AND COPPER IN THE DESIGN OF A TRANSFORMER

The previous section illustrated the fact that reducing a winding conductor's length enables a corresponding reduction in the conductor's cross-sectional area to maintain the same total I^2R losses. Therefore, to maintain constant losses, the required volume of copper is proportional to the square of the conductor length L.

$$V_{Cu} \propto L^2 \tag{2.10.1}$$

The required number of turns in a winding N is inversely proportional to the volts per turn generated by the core. The volts per turn are proportional to the total magnetic flux, and the flux is proportional to the cross-sectional area of the core A_{Fe} for a given allowable peak flux density, expressed as volts per turn.

$$N \propto \frac{1}{A_{Fe}} \tag{2.10.2}$$

From simple geometry, we know that the conductor's length is equal to the number of turns times the circumference of the coil. If the cross section of the core is nearly circular and the winding is placed directly over the core, the circumference of the coil is roughly proportional to the square root of the core's cross-sectional area.

$$L_{Cu} \propto N \times (A_{Fe})^{1/2} \propto (A_{Fe})^{-1/2} \tag{2.10.3}$$

$$V_{\text{Cu}} \propto L^2 \propto \frac{1}{A_{\text{Fe}}} \tag{2.10.4}$$

Assuming that the core's volume is roughly proportional to the core's cross-sectional area,

$$V_{\text{Cu}} \propto \frac{1}{V_{\text{Fe}}} \tag{2.10.5}$$

The relationships given in Eq. (2.10.5) indicate that the volume of copper required to limit I^2R losses is inversely proportional to the volume of the core for a given KVA rating, winding configuration, and applied voltage. In other words, adding 25% more core steel should permit a 25% reduction in the quantity of copper used in a transformer. This results in a 1:1 trade-off in copper volume vs. core volume.

Remember from Eq. (1.8.7), however, that the total core losses are proportional to the core *volume* for a given flux density. For example, if we decide to reduce the volume of copper by 25% by increasing the volume of core steel by 25%, the core losses will increase by 25% even though the conductor losses remain constant. In order to maintain the same core losses, the flux density must be reduced by increasing the cross-sectional area of the core, meaning that additional iron must be added. Therefore, the 1:1 trade off in copper volume vs. core volume is only a very rough approximation. There are also other practical physical limitations in selecting the dimensions of the core and windings; however, this exercise does illustrate the kinds of trade-offs that a transformer design engineer can use to optimize economy.

2.11 CONNECTING THREE-PHASE BANKS USING SINGLE-PHASE TRANSFORMERS

There can be advantages to using single-phase transformers to make a three-phase bank instead of building a three-phase unit. For instance, it may be impossible or impractical to fabricate or ship a three-phase transformer with an extremely large MVA capacity. A bank of three single-phase transformers may then be the solution, although the total size, weight, and cost of three single-phase units will probably exceed the size, weight, and cost of one three-phase unit. An additional advantage of the bank arrangement is that a failure of one single-phase unit will usually be less costly to repair than a failure of a larger three-phase unit. Furthermore, one spare single-phase transformer is usually all that is required to assure sufficient reliability for the entire bank. With a three-phase transformer, an additional spare three-phase transformer would be required, so the total cost of the installation plus a spare transformer is twice the cost of the installation alone. The total cost of a bank of single-phase transformers plus a spare is only 133% the cost of the bank alone. Therefore, the total cost of a bank of single-phase transformers plus a spare is probably less than the cost of a three-phase transformer plus a spare.

Either Y or Δ connections are possible with single-phase transformers connected in banks. It is extremely important that the single-phase transformers are carefully matched when they are banked together, especially when the Δ connection is used. Using mismatched transformers in the Δ connection will result in excessive circulating currents that will severely de-rate the bank or cause overheating. The operation of mismatched transformers will be explored further in Chapter 3.

One interesting configuration for a three-phase bank is the open Y-Δ connection used extensively in rural distribution systems. The open Y-Δ connection uses two single-phase transformers, as shown in Figure 2.16. An open Y-Δ connection requires only two phases plus the neutral on the primary side of the bank in order to develop a three-phase voltage at the secondary. This is an obvious cost saving (in addition to the avoided cost of a third transformer) when the installation is far away from a three-phase primary circuit. If one of the transformers is center-tapped as shown in Figure 2.16, then the bank provides a single-phase lighting leg in addition to a three-phase power circuit.

Example 2.3

An open Y-Δ bank comprised of two 10 KVA single-phase distribution transformers supplies a balanced 240 V three-phase load. What is the maximum KVA that can be supplied without exceeding the rating of the transformers?

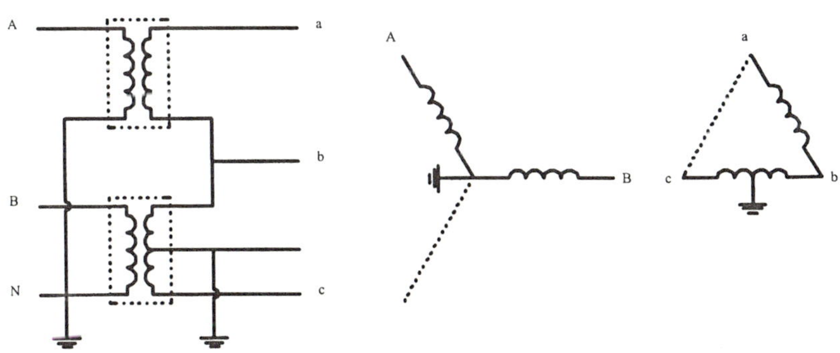

FIGURE 2.16 Open Y-Δ transformer bank with two single-phase transformers supplying a three-phase load.

Suppose we connect a 20 KVA load, which is equal to the total KVA rating of the two transformers. The phase currents are calculated as follows:

$$i = \frac{1/3 \times 20{,}000 \text{ VA}}{240 \text{ V} \times 0.577} = 48.1 \text{ A}$$

All of the phase current flows through the winding since the open ends of the Δ are connected to the load. The KVA transformed by each transformer is equal to the secondary voltage times the phase current.

$$\text{KVA transformed} = 240 \text{ V} \times 48.1 \text{ A} = 11.544 \text{ KVA}$$

In order not to exceed the 10 KVA transformer rating, the load must be reduced to 86.6% of 20 KVA. This demonstrates one disadvantage of the open Y-Δ bank: The available KVA capacity is only 86.6% of the total KVA rating of the two transformers. In practice, transformers used in open Y-Δ banks are sized so that a third transformer can be added at a later date when the customer's load increases beyond the 86.6% limit to provide a full-capacity Y-Δ bank.

2.12 TRANSFORMING THREE-PHASE VOLTAGES INTO TWO-PHASE VOLTAGES

Occasionally, although rarely, one still may encounter a two-phase power system that is supplied by a three-phase source. Two-phase systems can have three-wire, four-wire, or five-wire circuits. Note that a two-phase system is *not* merely two-thirds of a three-phase system. Balanced three-wire, two-phase circuits have two phase wires, both carrying approximately the same amount of current, with a neutral wire carrying 1.414 times the currents in the phase wires. The phase-to-neutral voltages are 90° out of phase with each other. Four-wire circuits are essentially just two ungrounded single-phase circuits that are electrically 90° out of phase with each other. Five-wire circuits have four phase wires plus a neutral; the four phase wires are 90° out of phase with each other.

The easiest way to transform three-phase voltages into two-phase voltages is with two conventional single-phase transformers. The first transformer is connected phase-to-neutral on the primary (three-phase) side and the second transformer is connected between the other two phases on the primary side. The secondary windings of the two transformers are then connected to the two-phase circuit. The phase-to-neutral primary voltage is 90° out of phase with the phase-to-phase primary voltage, producing a two-phase voltage across the secondary windings. This simple connection, called the T connec-

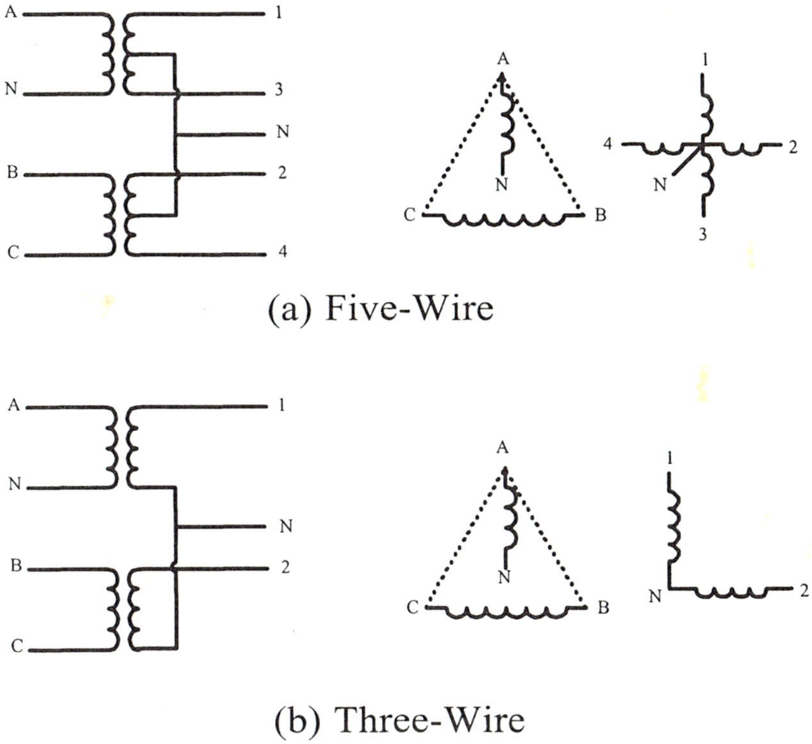

FIGURE 2.17 T transformer connection between a three-phase system and a two-phase system.

tion, is shown in Figure 2.17. The main advantage of the T connection is that it uses transformers with standard primary and secondary voltages. The disadvantage of the T connection is that a balanced two-phase load still produces unbalanced three-phase currents; i.e., the phase currents in the three-phase system do not have equal magnitudes, their phase angles are not 120° apart, and there is a considerable amount of neutral current that must be returned to the source.

2.13 THE SCOTT TRANSFORMER CONNECTION

In order to overcome the disadvantage of the T connection, the Scott connection uses two single-phase transformers of a special design to transform three-phase voltages and currents into two-phase voltages and currents. The first transformer, called the "main," has a center-tapped primary winding connected to the three-phase circuit with the secondary winding connected to the two-phase circuit. It is vital that the two halves of the center-tapped primary winding are wound around the same core leg so that the ampere-turns of the two halves cancel out each other. The ends of the center-tapped main primary winding are connected to two of the phases of the three-phase circuit.

The second transformer, called the "teaser," has one end of its primary winding connected to the third phase of the three-phase circuit and the other end connected to the center tap of the primary winding of the main. The Scott connection requires no primary neutral connection, so zero-sequence currents are blocked. The secondary windings of both the main and teaser transformers are connected to the two-phase circuit. The Scott connection is shown in Figure 2.18 for a two-phase, five-wire circuit, where both secondary windings are center-tapped and the center taps are connected to the neutral of the five-wire circuit. Three-wire and four-wire configurations are also possible.

If the main transformer has a turns ratio of 1 : 1, then the teaser transformer requires a turns ratio of 0.866 : 1 for balanced operation. The principle of operation of the Scott connection can be most easily seen by first applying a current to the teaser secondary windings, and then applying a current to the main secondary winding, calculating the primary currents separately and superimposing the results.

Apply a 1.0 per unit load connected between phase 1 and phase 3 of the secondary:

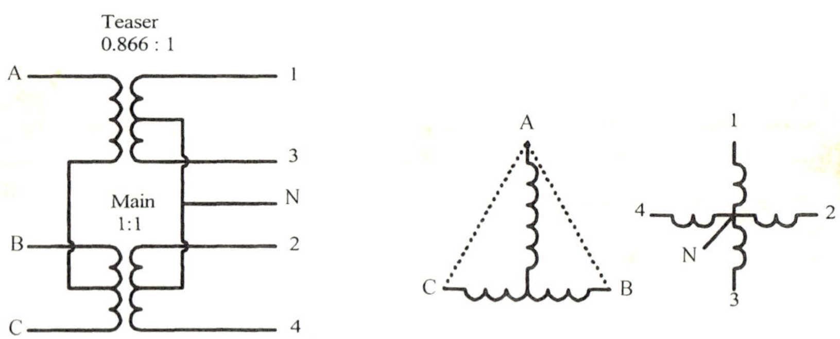

FIGURE 2.18 Scott transformer connection between a three-phase system and a two-phase system.

Secondary current from the teaser winding into phase 1 $= 1.0\angle 90°$

Secondary current from the teaser winding into phase 3 $= -1.0\angle 90°$

Primary current from A phase into the teaser winding $= 1.1547\angle 90°$

Primary current from B phase into the main winding $= 0.5774\angle 90°$

Primary current from C phase into the main winding $= -0.5774\angle 90°$

The reason that the primary current from A phase into the teaser winding is 1.1547 per unit is due to 0.866:1 turns ratio of the teaser, transforming 1/0.866 = 1.1547 times the secondary current. This current *must* split in half at the center tap of the main primary winding because both halves of the main primary winding are wound on the same core and the total ampere-turns of the main winding must equal zero.

Apply a 1.0 per unit load connected between phase 2 and phase 4 of the secondary:

Secondary current from the main winding into phase 2 $= 1.0\angle 0°$

Secondary current from the main winding into phase 4 $= -1.0\angle 0°$

Primary current from B phase into the main winding $= 1.0\angle 0°$

Primary current from C phase into the main winding $= -1.0\angle 0°$

Primary current from A phase into the teaser winding $= 0$

Superimpose the two sets of primary currents:

$$I_a = 1.1547\angle 90° + 0 = 1.1547\angle 90°$$

$$I_b = 0.5774\angle 90° + 1.0\angle 0° = 1.1547\angle -30°$$

$$I_c = -0.5774\angle 90° - 1.0\angle 0° = 1.1547\angle 210°$$

Notice that the primary three-phase currents are balanced; i.e., the phase currents have the same magnitude and their phase angles are 120° apart. The apparent power supplied by the main transformer is greater than the apparent power supplied by the teaser transformer. This is easily verified by observing that the primary currents in both transformers have the same magnitude; however, the primary voltage of the teaser transformer is only 86.6% as great as the primary voltage of the main transformer. Therefore, the teaser transforms only 86.6% of the apparent power transformed by the main.

We also observe that while the total *real* power delivered to the two-phase load is equal to the total *real* power supplied from the three-phase system, the total *apparent* power transformed by both transformers is greater than the total *apparent* power delivered to the two-phase load. Using the numerical example above, the total load is 2.0 per unit. The apparent power transformed by the teaser is $0.866 \times I_a = 1.0$ per unit, and the apparent power transformed by the main is $1.0 \times I_b = 1.1547$ per unit for a total of 2.1547 per unit of apparent power transformed. The additional 0.1547 per unit of apparent power is due to parasitic reactive power flowing between the two halves of the primary winding in the main transformer.

Single-phase transformers used in the Scott connection are specialty items that are virtually impossible to buy ''off the shelf'' nowadays. In an emergency, standard distribution transformers can be used, as illustrated in the following example.

Example 2.4

(*Note: The following example is taken from an actual case that occurred a number of years ago.*)

A small industrial electric customer has a plant with an antiquated two-phase electrical system. The customer has a Scott-connected transformer bank connecting this system to the local utility company which provides power. This Scott-connected bank catastrophically failed and there are no replacement transformers readily available. The utility company was called in to restore service to the two-phase system, but the utility only has standard single-phase and three-phase distribution transformers in its storerooms. The customer has a five-wire, two-phase system with a phase-to-neutral voltage of 120 V. The utility company's distribution system has a phase-to-phase voltage of 12,470 V and a phase-to-neutral voltage of 7200 V. Customers are routinely supplied with 120/240 V service using single-phase transformers connected either phase-to-phase or phase-to-neutral on the primary side.

One solution would be to use the T connection with the standard transformers that the utility had on hand, but this would result in unbalanced three-phase currents. A bright young engineer working for the utility found a way to construct a Scott-connected bank using standard transformers.

The teaser was replaced by a standard distribution transformer with a 12,470 V primary and a 120/240 V center-tapped secondary. The main transformer was replaced by two standard distribution transformers, each having a 7200 V primary and a 120/240 V center-tapped secondary. The primary windings of the replacement main transformers were connected in series and their secondary windings were connected in parallel. These connections are shown in Figure 2.19.

Since the two transformers substituted for the main transformer are not wound on the same core leg, the secondary windings of the main transformers

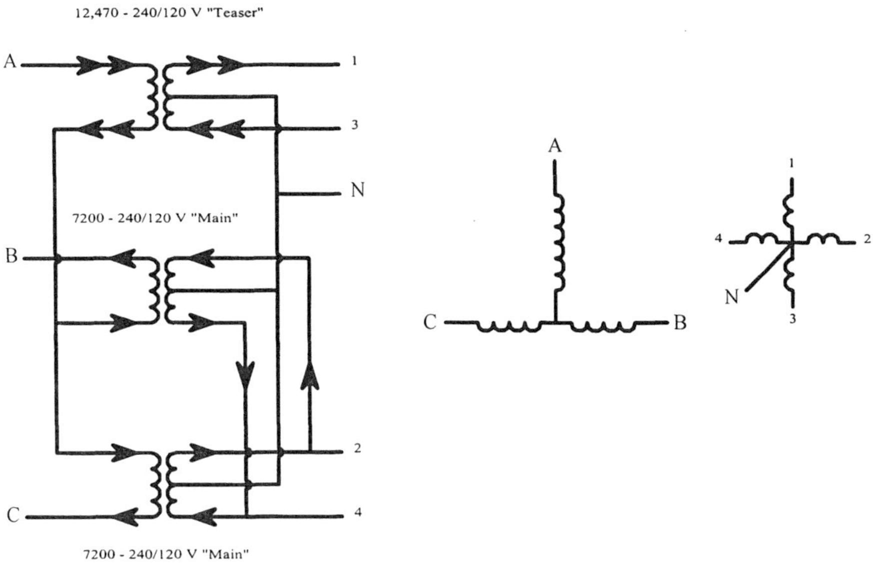

FIGURE 2.19 Scott-connected transformer bank using three standard single-phase distribution transformers, showing the current in the teaser primary current splitting evenly between the two main primaries.

are interconnected to force the primary current in the primary winding of the teaser to split evenly between the two primary windings of the main in the same manner that a center-tapped winding forces the teaser current to split evenly in the primary winding of the main in a Scott connection.

For a five-wire, two-phase circuit, the center tap of each of the three secondary windings is connected to the common neutral. For a four-wire, two-phase circuit, the center taps of the secondary windings are not brought out.

Note that the voltage across the 12,470 V primary winding of the teaser is 86.6% of the voltage across the two 7200 V primary windings in series of the main, just like a Scott-connected transformer. This produces balanced operation; however, when the bank is connected to a 12,470 V three-phase source, the phase-to-neutral output voltage is only 86.6% of 120 V, or 104 V. Fortunately in this case, the three-phase distribution voltage at the customer's location was 5% higher than the nominal voltage and the transformers also had taps to boost the output voltage by another 5%. The combination of the 5% higher system voltage and the 5% tap boost provided a two-phase secondary voltage of about 115 V. This was still below the 120 V nominal voltage but it was within the operating voltage range of the customer's two-phase equipment.

2.14 THREE-PHASE TRANSFORMER DESIGNS

In the preceding sections we saw some of the ways that single-phase transformers can be connected in three-phase and two-phase systems. It is sometimes advantageous to construct a single three-phase transformer instead of using a bank of single-phase transformers. For instance, a three-phase transformer can often be more economical to build by enclosing one core and coil structure inside one transformer tank instead of building three separate core and coil structures and tanks.

The first basic design for three-phase transformers to be discussed is the core form design. This design includes the three-legged core form design and the five-legged core form design. In the three-legged core form design, three sets of windings are placed over three vertical core legs. Each core leg and its set of windings is corresponds to one phase. Each core leg is joined to a top yoke and a bottom yoke, which complete the magnetic circuit. The general layout of the three-legged core form design is shown in Figure 2.20. The set

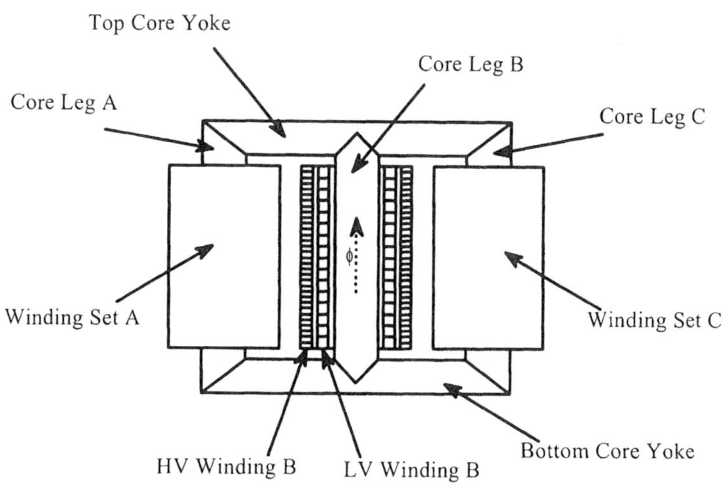

FIGURE 2.20 Three-legged core form transformer.

of windings on the center phase are cut away to show the placement of the LV and HV windings around the center core leg.

The construction of the core follows the same principles as described in Chapter 1. The core is made of thin steel laminations that are stacked together. The cross section area of the yokes must be sufficient to carry the full flux of each phase. There are various methods of making joints between the core legs and the yokes; the method favored at the present is the mitered joint used with the modern grain-oriented steels. The core and coil cross sections are usually circular, although a square design is sometimes used. Any of the winding connections described earlier, such as Y, Δ, and zigzag, can be implemented in the three-legged core form design.

The one feature that makes the three-legged core form design unique is the fact that there is apparently no return path for the magnetic flux converging in the top and bottom yokes from the three core legs. This would imply that the 3 fluxes arriving at the top core yoke would have to jump across a huge air gap to return to the bottom core yoke. Further examination reveals the secret: The flux through each core leg is sinusoidal since the voltages applied across the windings are sinusoidal. As long as fluxes have the same magnitude and their phase angles are electrically 120° apart, the fluxes will cancel out each other in the top and bottom yokes. Therefore, for three-phase balanced applied voltages, no flux return path is required.

If the applied voltages are not balanced, however, then there will be residual flux that must be dealt with. Since the unbalanced flux travels along a high-reluctance path through a very long air gap and since the inductance of a coil is inversely proportional to the reluctance of its flux path, the unbalanced voltage component encounters only a small inductance. Therefore, the low inductance presented by the return path of a three-legged core transformer tends to "short out" any voltage unbalance. This effectively links the three phases together through a mutual magnetic path and tends to reduce voltage unbalance between phases. We shall see in a later chapter that this has the same effect as a Δ-connected *tertiary winding*.

This has a major advantage in the case of the Y-Y connection. We saw earlier that if the neutrals are not connected to an external circuit, then third harmonic voltages would be impressed across the windings because third harmonic magnetizing currents cannot flow. In a three-legged core design, the third harmonic flux from the three core legs add in phase at the top and bottom yokes and encounter a high reluctance path. Restricting the third harmonic flux greatly reduces the third harmonic voltages that would otherwise exist in Y-Y transformers. For this reason, along with the economic savings of not having to provide a magnetic return path between the top and bottom yokes, three-legged core form designs are used extensively with Y-Y transformers.

In large power transformers, third harmonic flux and stray 60 Hz flux from unbalanced voltages may leave the iron core and enter free space inside the transformer. This induces currents in the internal metal parts of the transformer and may cause severe localized overheating. The *five-legged core form* design solves this problem by providing flux paths around the three core legs between the top and bottom yokes. The layout of the five-legged core form transformer is shown in Figure 2.21. Residual flux, labeled ϕ, is the total flux arriving at the top core yoke from the core legs of the three phases. A fourth and fifth core leg provide two return paths for the residual flux from the top to the bottom yoke.

The top and bottom core yokes are often made with a reduced cross-sectional area because they do not have to carry the full complement of flux from each phase. This has the advantage of reducing the overall height of the transformer. The disadvantage of reducing the cross-sectional area of the yokes is that a portion of the normal flux from the outer phases must now flow through the fourth and fifth core legs. This makes the actual flux path uncertain and makes calculating core losses difficult.

The five-legged core form design has electrical characteristics that are very similar to a bank of three single-phase transformers; i.e., the three phases operate more or less independently with a relatively small magnetic interaction between the phases.

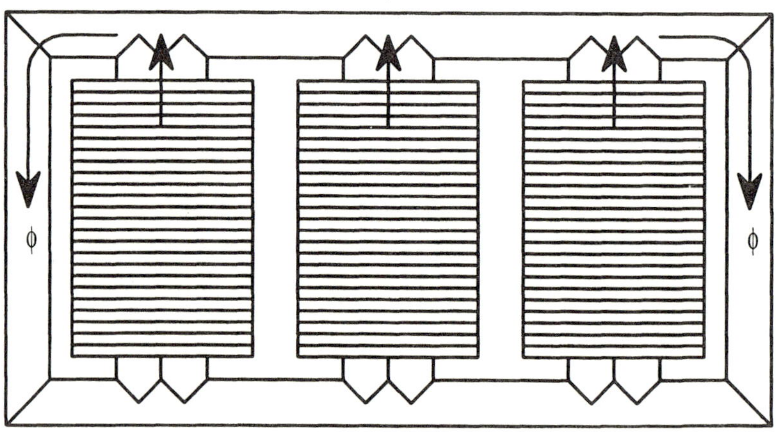

FIGURE 2.21 Five-legged core form transformer.

The *shell form design* is a completely different design from the three-phase core form design. In a shell form transformer, the windings are constructed from flat coiled spirals that are stacked together like pancakes. For this reason, the windings in a shell form design are often referred to as *pancake windings*. In a two-winding shell form transformer, the low-voltage winding is usually split into two windings with the high-voltage winding sandwiched between the two halves of the low-voltage winding. Instead of a circular shape, the pancake coils are actually square shaped with the outer corners rounded off.

The centers of the pancake coils are hollow and square-shaped. Core legs with square cross sections pass through the centers of the pancake coils. The core legs are laid horizontally so the coils are stacked horizontally on edge. The return paths for the core go around the coils forming a "shell" around the windings; hence the name *shell form*. The basic shell form design is shown schematically in Figure 2.22.

The core without windings is shown on the left hand part of Figure 2.22. The core is divided into two halves with each half carrying approximately 50% of the total flux. The three sets of windings are placed around the longitudinal sections of the core as shown in the middle part of the figure as viewed

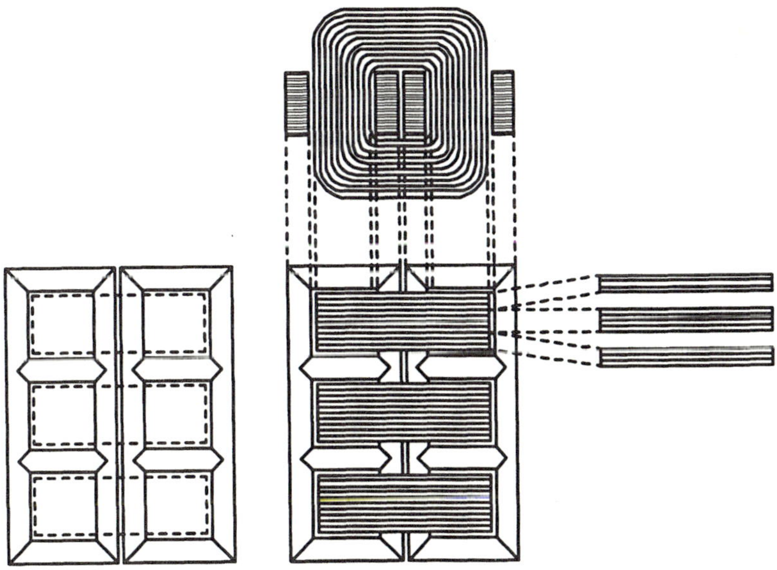

FIGURE 2.22 Three-phase shell form transformer.

from above the transformer. The top part of the figure shows the end cut away, clearly illustrating that the pancake coils are square-shaped with rounded corners. The edges of the core laminations are also seen in this view. The right part of the figure is an exploded end view of the pancake windings divided into three sections: The low-voltage winding is split into two halves, which sandwich the high-voltage winding in the middle.

The electrical characteristics of the shell form transformer are similar to a five-legged core form transformer or a bank of three single-phase transformers; i.e., there is not a strong magnetic interaction among the different phases at normal levels of excitation. All the various three-phase winding connections are available in three-phase shell form transformers, such as the Y, Δ, and zigzag. The major advantages of the shell form design are that it is more compact than core form transformers and it has great mechanical strength. The main disadvantages of the shell form design are that it is much more difficult to assemble than the core form design, and there are tighter electrical clearances to contend with.

The shell form transformer technology was developed by Westinghouse Electric Corporation and was licensed to a number of transformer manufacturers. Both single-phase and three-phase shell form transformers are made. There are very few manufacturers that still make shell form transformers today, despite the fact that this design has definite advantages in certain applications.

Example 2.5

In a shell form transformer, the cross-sectional areas of each of the core sections are only sufficient to carry a peak flux density equal to the flux produced by one set of phase windings. Sketch a diagram of a three-phase shell form transformer core showing the directions that the flux paths must take in order that the total flux in any section does not exceed the magnitude of flux generated by one phase. There are three components of flux that are electrically 120° apart. Label these flux components 1, 2, and 3. Sketch the flux throughout the core, indicating the directions of the flux in each core section.

The solution appears in Figure 2.23. Assuming that the flux generated by each phase is sinusoidal and the three phases are electrically 120° apart, the flux components cancel at each point marked by the letter X. The flux components flowing in the lateral sections of the core are not generated directly by the windings but are the fluxes of two phases combining at the points marked X.

Note that the direction of flux in the longitudinal core leg in the middle

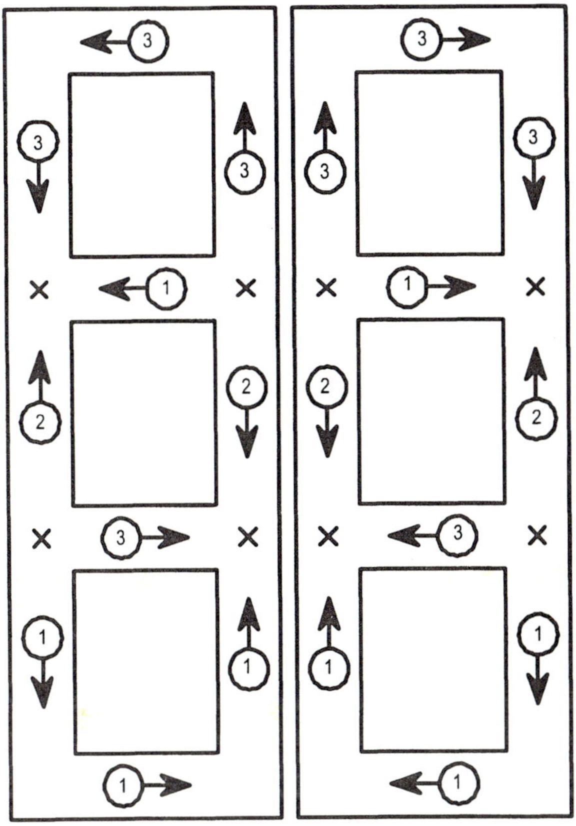

FIGURE 2.23 Flux paths in a three-phase shell form transformer.

phase must be in a direction that is *opposite* from the direction of the flux in the core legs of the other phases in order to cancel out the flux arriving at each of the points marked X. If the directions of the flux in all the three core legs are in the same direction, then the magnitude of the total flux arriving at each of the points marked X will be 1.73 times the magnitude of the flux produced by one phase. This will drive the lateral core sections into saturation.

The desired configuration of the core flux shown in Figure 2.23 is provided simply by connecting the center-phase windings with polarity opposite of the other two phase windings.

2.15 STANDARD TERMINAL MARKINGS FOR TRANSFORMERS

The electrical connections to transformer windings are brought out of the transformer enclosure by leads that pass through insulated *bushings*, which will be covered in more detail in Chapter 6. The bushings can be located on the top or on the sides of the transformer and are marked in a standard way that is consistent throughout the industry [2].

For single-phase two-winding transformers, the high-voltage bushings are labeled H1 and H2 and the low-voltage bushings are labeled X1 and X2. The labels indicate the polarity of the terminals; i.e., if the voltage at the H1 bushing is instantaneously positive with respect to H2, then the voltage at the X1 bushing is instantaneously positive with respect to X2. If the windings of a single-phase transformer are connected line-to-neutral, then the H1 and X1 bushings are usually connected to the line with the H2 and X2 bushings are connected to the neutral.

The physical placement of the bushings on the transformer is also standardized. If the bushings are mounted on the top of the transformer, the H1 bushing is always located in the upper-left quadrant of the transformer lid as seen from above the transformer, and the H2 bushing is located in the upper-right quadrant. The locations of the X1 and X2 bushings can either be lower left and lower right, respectively, or lower right and lower left, respectively. In the former case the transformer is said to have a *subtractive polarity*, and in the latter case the transformer is said to have an *additive polarity*.

The way to remember the convention for additive and subtractive polarity is to draw a voltage vector arrow from the H2 bushing to the H1 bushing and draw a voltage vector arrow from the X2 bushing to the X1 bushing, indicating an instantaneous positive polarity at the H1 and X1 terminals. If the voltage vector arrows point in the *same* direction in a circular path around the perimeter of the transformer, the transformer is additive. If the vector arrows point in *opposite* directions in a circular path around the perimeter of the transformer, the transformer is subtractive. Schematic diagrams of additive and subtractive polarity transformers are shown in Figure 2.24.

The polarity of an unmarked single-phase transformer can be determined by a simple test: Apply 120 VAC across the high-voltage bushings (*not across the low-voltage bushings, please*!). Connect a jumper wire from one of the high-voltage bushings to the nearest low-voltage bushing on the opposite side of the transformer. Measure the voltage between the *other* high-voltage bushing and the *other* low-voltage bushing. If the measured voltage is less than 120 V, then the transformer is subtractive. If the measured voltage is greater

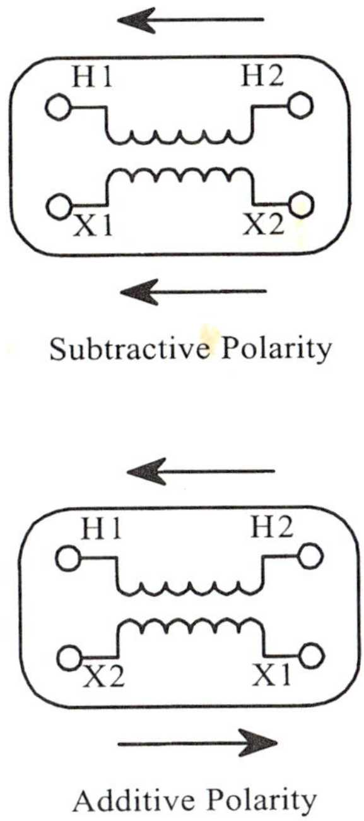

FIGURE 2.24 Placement of the bushings for additive-polarity and subtractive-polarity single-phase transformers.

than 120 V, then the transformer is additive. A method of verifying transformer polarity that is more safe uses a transformer turns ratio (TTR) test set, which will be discussed in Chapter 8.

Single-phase transformers with ratings 200 KVA and below and having high-voltage ratings of 8660 V and below are generally additive polarity, whereas single-phase transformers above 200 KVA or with high-voltage ratings above 8660V are generally subtractive polarity [3].

Although this is the standard convention, it may be possible for two transformers with the same KVA and voltage rating to have opposite polarities. Note that nothing prevents connecting subtractive and additive transformers together in a three-phase bank as long as the correct polarities of the *leads* are observed. There are no electrical differences between a subtractive and an additive transformer; only the physical locations of the low-voltage bushings are different.

If a low-voltage winding is center tapped, which is common in single-phase distribution transformers, then the low-voltage bushings are labeled X1, X2, and X3. The X1 and X3 bushings are connected across the full winding and the X2 bushing is connected to the center tap. The instantaneous polarity of voltage at the X1 bushing is the same as the polarity of the voltage at the H1 bushing.

Polarity dots are sometimes used instead of labeling the transformer leads H1 and X1 labels, particularly with instrument transformers. A polarity dot on a high-voltage lead is equivalent to marking the lead H1, and a polarity dot on a low-voltage lead is equivalent to marking the lead X1. In other words, polarity dots indicate the leads that have the same instantaneous voltage polarity.

Current transformers are instrument-type transformers that are used to transform currents instead of transforming voltages. For a current transformer, current flowing *into* a terminal, lead, or window that is marked with a polarity dot produces an output current flowing *out of* the terminal or lead that is marked with a polarity dot on the other winding.

The standard terminal markings for three-phase transformers are only slightly more complicated than markings for single-phase transformers. For three-phase, two-winding transformers, the phase leads of the high-voltage

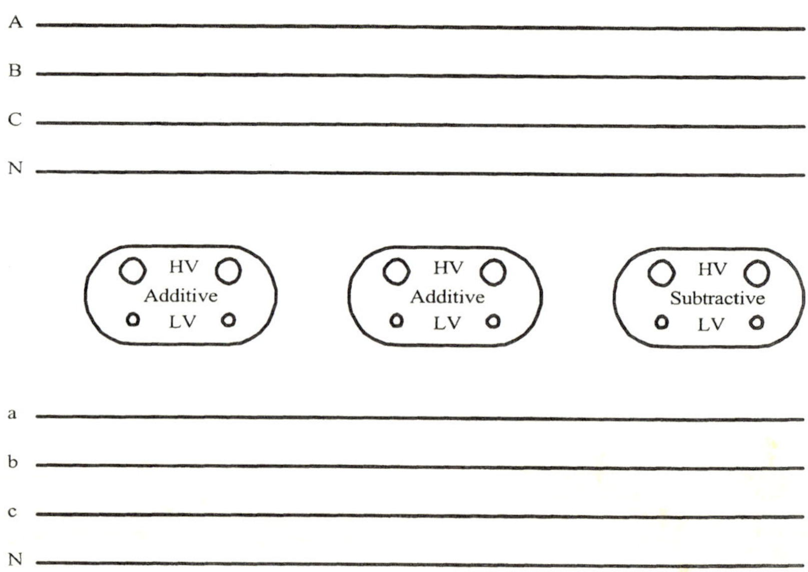

FIGURE 2.25 Problem of Example 2.6.

windings are connected to high-voltage bushings labeled H1, H2, and H3. If the high-voltage winding has an external neutral, this is brought out to a neutral bushing marked H0. The phase leads of the low-voltage windings are connected to low-voltage bushings labeled X1, X2, and X3. If the low-voltage winding has an external neutral, this is brought out to a neutral bushing marked X0. There is no H0 or X0 bushing for a Δ winding or a Y winding without an external neutral connection.

If the high-voltage bushings are connected to a three-phase system with a phase rotation H1-H2-H3, the phase rotation of the low-voltage bushings is X1-X2-X3. In standard Y-Y and Δ-Δ connected transformers, the voltage at the X1 bushing is in phase with the voltage at the H1 bushing. In standard Y-Δ and Δ-Y transformers connected to a system with a 1-2-3 phase rotation, the X1 voltages *lags* the H1 voltage by 30°.

Nonstandard phase displacements can also be specified for transformers used in special applications. For example, Y-Y transformers can be specified with 180° phase displacement between the H1 and X1 terminals instead of a 0° phase displacement, and Δ-Y transformers can have the low-voltage phases leading the high-voltage phases by 30° for a 1-2-3 phase rotation instead of lagging the high-voltage phases by 30°.

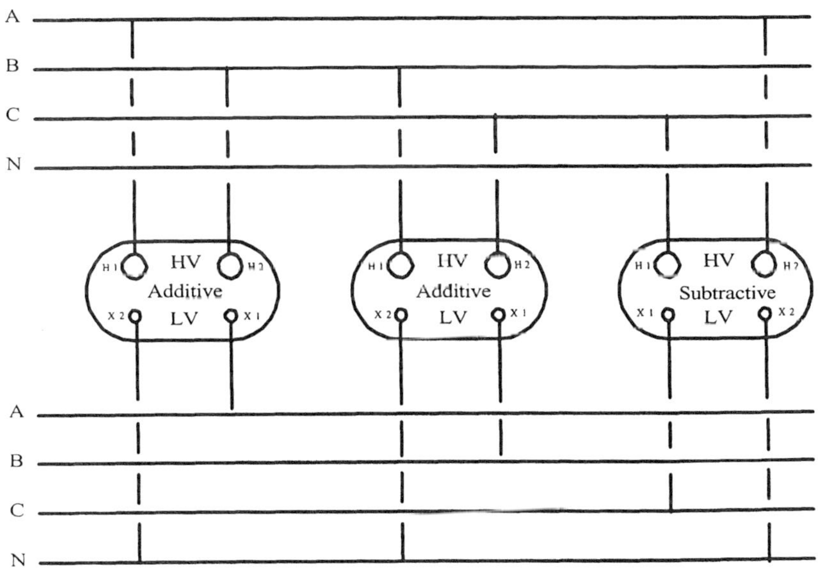

FIGURE 2.26 Solution of Example 2.6.

FIGURE 2.27 Three-phase core form transformer in assembly at the factory.

FIGURE 2.28 Three-phase shell form transformer in assembly at the factory.

The placement of the bushings in a three-phase transformer is also standardized. When bushings are mounted on the top of the transformer lid, the high-voltage bushings are arranged H0, H1, H2, and H3 from left to right in the upper half of the lid, as seen from above the transformer. The low-voltage bushings are arranged X0, X1, X2, and X3 from left to right in the lower half of the lid, as seen from above the transformer. If side-mounted low-voltage bushings are used, they are arranged X0, X1, X2, and X3 from left to right facing the transformer from the low-voltage side. If side-mounted high-voltage bushings are used, they are arranged H3, H2, H1 and H0 from left to right facing the transformer from the high-voltage side. The terms ''additive polarity'' and ''subtractive polarity'' do not apply to three-phase transformers.

Example 2.6

Assume that the single-phase transformers shown in Figure 2.25 have standard terminal markings. Label the terminals and show the proper connections for a Δ-Grd.Y connection (with the low-voltage circuit Y-connected) with the high voltage *lagging* the low voltage by 30°.

The proper connections are shown in Figure 2.26. The H1 bushings are always on the upper left of the lid, per the standard arrangement. The low-voltage bushings are labeled according to whether the transformer is additive or subtractive. When the low-voltage windings are connected phase-to-neutral, the X1 bushings are normally connected to the phase wires and the X2 bushings are connected to the neutral. With the connection shown in the figure, the voltages at the high-voltage bushings lag the voltages at the low-voltage bushings by 30°.

Figure 2.27 shows the core and coil assembly of a three-phase core form tansformer seen from the high-voltage side of the transformer. Figure 2.28 shows the core and coil assembly of a three-phase shell form transformer.

REFERENCES

1. R. L. Bean, N. Chakan, Jr., H. R. Moore, and E. C. Wentz. *Transformers for the Electric Power Industry*. McGraw-Hill, New York, 1959, p. 99.
2. ANSI Std. C57.12.70–1978. American National Standard Terminal Markings and Connections for Distribution and Power Transformers. American National Standards Institute, New York, 1978, pp. 18–19.
3. IEEE Std. C57.12.00–2000. IEEE Standard General Requirements for Liquid-Immersed Distribution, Power, and Regulating Transformers. Institute of Electrical and Electronics Engineers, Inc., New York, 2000, pp. 11–12.

3

Transformer Impedance and Losses

Until now, we have considered only ideal, lossless transformers obeying the fundamental transformer laws: Voltages and currents are transformed strictly according to the transformer turns ratio, and the power output from the transformer is equal to the power input to the transformer. The conditions expressed by the ideal transformer laws are approached, but never realized in physical transformers. Transformed voltages and currents are always less than the values predicted by the turns ratio because of losses. This chapter explains the origins of transformer losses and develops the concept of equivalent circuits of real transformers.

3.1 LEAKAGE FLUX AND LEAKAGE REACTANCE

In an ideal two-winding transformer excited by the primary winding, *all* of the magnetizing flux is within the core and both the primary and secondary windings are linked by the same flux. Figure 3.1 depicts the ideal transformer in a cutaway diagram of one phase of a two-winding core form transformer with load currents in the primary and secondary windings, assuming a load

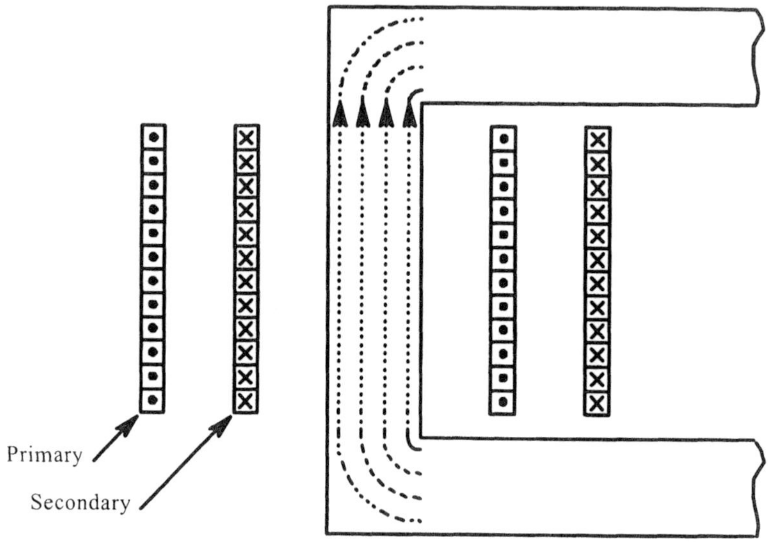

FIGURE 3.1 Flux in an ideal transformer under load. All of the magnetizing flux is contained in the iron core with the ampere-turns of the primary canceling the ampere-turns of the secondary.

is connected to the secondary winding and a voltage is applied across the primary winding.

In Figure 3.1, the currents flowing out of the page are denoted by the symbol "•" and currents flowing into the page are denoted by the symbol "×." Note that it is not necessary for the actual number of turns or the transformer turns ratio to be shown. Since the total ampere-turns in the primary are approximately equal to the total ampere-turns in the secondary, the windings are simply drawn with the same number of turns with the same current in each turn.

The magnetic flux is denoted by the dotted lines inside the core in Figure 3.1. For this particular transformer, it takes four lines of flux in the core in the upward direction to induce a voltage equal to the voltage applied across the primary winding. The spaces between the two windings and between the windings and the core are shown greatly exaggerated. The magnetizing current is assumed to be negligible compared to the load currents.

The situation in a real transformer is somewhat different than described above. The main difference is that all of the magnetic flux is not contained in the core. This is because the load currents in the primary and secondary windings are considerably greater than magnetizing current, so we cannot ignore the magnetic fields induced by these currents in the spaces surrounding the winding conductors. Figure 3.2 shows these magnetic fields as the dotted lines circulating around the windings in the spaces between the conductors.

The local magnetic flux circulating around the primary winding are approximately equal to the local magnetic flux circulating around the secondary winding because the primary ampere turns are approximately equal to the secondary ampere turns. Also, the magnetic flux is in phase with the currents that produce it. For the time being, suppose the load connected to the transformer secondary is purely inductive. The phase angles of both the load current and the circulating flux will lag the secondary voltage by 90°. Because induced voltages are proportional to the *rate of change* of flux, any flux pointing in the upward direction that is enclosed by the primary or secondary windings will induce voltages *in phase* with the applied voltage. Conversely, any flux pointing in the downward direction that is enclosed by the primary or secondary windings will induce voltages in these windings 180° *out of phase* with the applied voltage.

If the number of lines of flux in the core would remain constant at four lines pointing upward, the net number of flux lines contained pointing in the

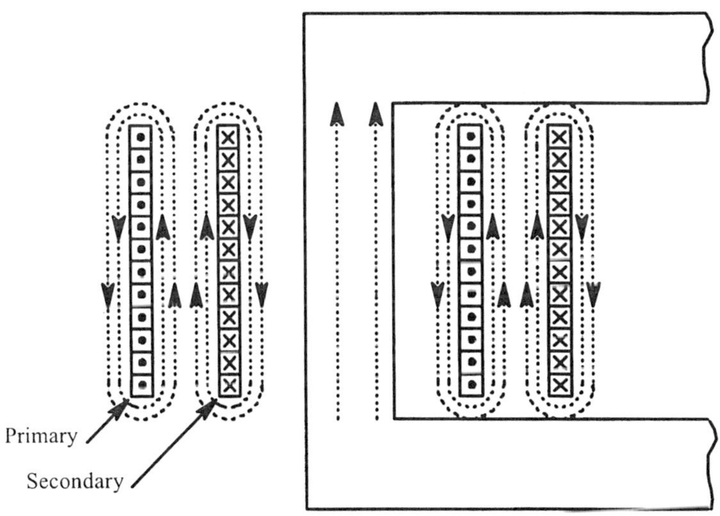

FIGURE 3.2 Flux in a real transformer under load. Flux generated by the conductor currents surround the windings, excluding flux from the core and reducing the secondary voltage.

upward direction enclosed by the primary winding would increase by two lines. This would naturally increase the voltage induced across the primary. In order for the voltage induced across the primary to remain equal the applied voltage across the primary, the net number of flux lines pointing in the upward direction must remain the same. This can be accomplished only if the number of lines within the core decreases to two lines pointing upward. Consequently, two lines of flux have been *excluded* from the core because of the flux circulating in the spaces around the windings.

The opposite situation exists for the secondary winding. If the number of lines of flux in the core would have remained constant at four lines pointing upward, the net number of flux lines pointing in the upward direction enclosed by the secondary winding would have decreased by two lines. But as shown in the preceding paragraph, the number of lines in the core must also decrease by two lines to balance the voltage across the primary winding. Therefore, the net number of flux lines in the vertical direction enclosed by the secondary winding will decrease by four lines, which means there are no lines enclosed by the secondary winding and the induced voltage across the secondary winding will be reduced to zero.

The above situation can actually exist when the secondary winding is short-circuited, so short-circuiting the secondary winding results in a purely inductive winding current with some of the flux excluded from the core. Note, however, that some residual flux remains in the core and that the short-circuit currents are still finite. This suggests that some sort of inherent restriction or impedance limits the short-circuit currents in the transformer.

The composite of all flux in the spaces around and between the conductors is called *leakage flux*, because this can be visualized as flux ''leaking'' out of the core through empty space. Because the permeability of iron is finite, there will always be some flux that leaks from the core, even without any load current. However, the leakage flux increases considerably as the currents in the windings increase. Since the leakage flux travels through empty space instead of through iron, there is no hysteresis or saturation and the permeability in this region is a constant equal to μ_0. Recalling that the ratio B/H is equal to the permeability and that H is directly proportional to the load current, the leakage flux is in phase with, and is directly proportional to, the load current:

$$\phi_{\text{leakage}} \propto B_{\text{leakage}} = \mu_0 \times H_{\text{leakage}} \propto I_{\text{load}} \tag{3.1.1}$$

The voltage induced across the secondary winding is equal to minus the rate of change of the flux in the core:

$$E_s = -\frac{d\phi_{\text{core}}}{dt} \tag{3.1.2}$$

Let

E_s = secondary voltage assuming no leakage flux
E'_s = secondary voltage with leakage flux

$$\Delta E_s = E_s - E'_s \tag{3.1.3}$$

As the magnitude of the leakage flux increases, the magnitude of flux in the core decreases. The drop in secondary voltage is therefore directly proportional to the rate of change of the load current:

$$\Delta E_s = \frac{d\phi_{\text{leakage}}}{dt} \propto dI_{\text{load}} \tag{3.1.4}$$

For a purely inductive load, ΔE_s is in phase with E'_s. For a purely resistive load, ΔE_s leads E'_s by 90°. The vector relationships are shown in Figure 3.3.

The relationship between load current and secondary voltage is equivalent to placing a physical impedance in series with the output of an ideal

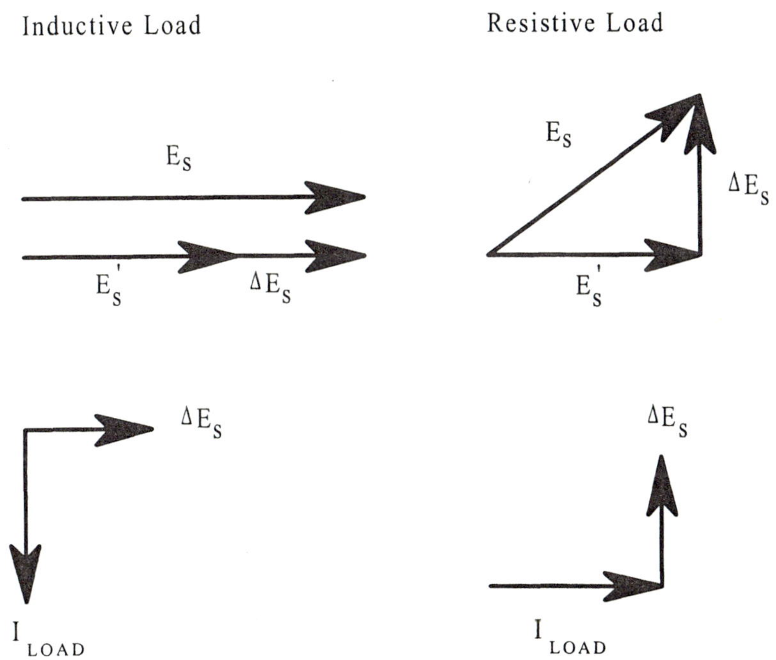

FIGURE 3.3 Vector relationships between the no-load secondary voltage, the secondary voltage with load, and load current.

transformer. This equivalent impedance is called the *leakage reactance* of the transformer because it is an external manifestation of the effect of leakage flux on secondary voltage. Leakage reactance limits short-circuit currents, and is therefore useful, but it also reduces the secondary voltage under load (regulation) and consumes reactive power. Varying the winding dimensions and the spacing between the windings can control the amount of leakage reactance. For example, if the spaces between the windings are kept small, there will be fewer flux lines contained in these spaces and the leakage reactance will be small. Placing the windings as close together as possible increases the "coupling" between them, which is the same thing as reducing the leakage reactance.

3.2 CONDUCTOR LOSSES

Chapter 2 briefly discussed I^2R conductor losses in transformer windings. It was determined that the resistance of the winding is proportional to the conductor length divided by the conductor cross-sectional area. While this is true for DC currents, it is actually an oversimplification for AC currents where additional conductor losses must be taken into account. These additional losses affect the transformer design to some extent.

When an alternating magnetic field is applied to any conductor, eddy currents are induced around the paths surrounding the lines of magnetic flux that penetrate the conductor. These currents generate local I^2R losses even if the conductor itself is not carrying any net electrical current.

Large amounts of leakage flux can occur when a transformer is heavily loaded. The magnetic fields associated with leakage flux not only penetrate the winding conductors themselves, but can involve other metallic parts as well. The eddy currents that are induced by these fields are proportional to the leakage flux, which in turn is proportional to the load currents. Therefore, the square of eddy currents and the eddy-current losses are both proportional to the square of the load current. These eddy losses are externally manifested by a component that increases the effective resistance of the conductors, even if the eddy losses occur in metallic parts that are electrically isolated from the conductors. Let this eddy-loss component of the conductor resistance be denoted R_e.

When an AC current flows in a conductor, the magnetic fields within the conductor form a series of concentric circles. The flux density B at any point in the conductor is proportional to the total current enclosed by the magnetic path divided by the length of the circular path. Moving away from the center of the conductor, the total current enclosed by the path tends to increase faster than the length of the path. Therefore, the flux density increases near the outer edges of the conductor. The direction of the magnetic field is perpendicular to the direction of the current flow, and this forces current toward the edge of the conductor and reduces the flux density near the center at the same time. The concentration of current toward the edge of a conductor is called the *skin effect*, reducing the area of the conductor that actually carries current and increasing the effective resistance of the conductor. The skin effect is more pronounced for large-diameter conductors. Let the skin-effect component of the conductor resistance be denoted R_s.

The total AC resistance of the conductor, including the eddy-loss component and the skin-effect component is expressed by the following equation:

$$R_{\mathrm{AC}} = R_{\mathrm{DC}} + R_e + R_s \tag{3.2.1}$$

where R_{AC} is the AC resistance of the conductor and R_{DC} is its DC resistance.

The conductor losses are equivalent to placing a lumped resistance in series with the terminals of an ideal transformer. Conductor losses are commonly referred to as load losses, because they result only from load currents. Load losses are sometimes referred to as *copper losses*; however, this is somewhat of a misnomer. As we have just seen, eddy-current losses in *any* metallic part that is exposed to leakage flux will still show up as load losses.

Load losses limit the KVA capacity of a transformer because the heat generated by these losses increase temperatures. Therefore, it is highly desirable to reduce the load losses as much as possible by reducing the AC resistance of the conductor. Reducing R_{DC} as well as R_e and R_s can do this. Reducing R_{DC} can be done by shortening the conductor length and/or by increasing the conductor cross-sectional area. Shortening the conductor length can only be achieved to a point, and increasing the conductor cross-sectional area has the unfortunate effect of increasing both the eddy-current losses and the skin-effect losses. These losses can be reduced by special conductor designs.

Subdividing the conductors into strands that are insulated from each other to break up the eddy current paths can reduce eddy-current and skin-effect losses. Generally, the strands have a rectangular shape with the long dimension oriented in the same direction as the leakage flux. By subdividing one large-area conductor into a number of small-area conductors, the skin effect is substantially reduced as well.

If a number of strands were simply bundled together, the strands would be exposed to varying amounts of leakage flux and different induced voltages. This would force more current to flow in some strands and less in others, increasing the total losses. If the strands are properly *transposed*, each strand is exposed to the same amount of leakage flux with equal voltages induced

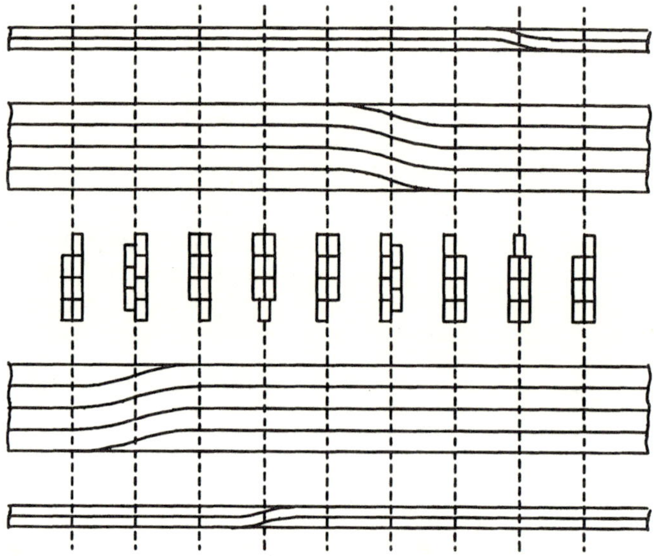

FIGURE 3.4 Seven-conductor transposed cable used in a transformer winding making a single transposition. After seven such transpositions, the conductors return to the original configuration.

in each of the strands. Transposition allows the current to divide equally among the strands to optimize the use of the entire cross section area of the conductor. The transposition of a seven-strand cable is illustrated in Figure 3.4, showing the top, front, end, back and bottom views in sequence. After seven complete transpositions, all the strands return to their original positions.

3.3 NO-LOAD LOSSES

Core losses were briefly discussed in Chapter 1. Alternating magnetic flux produces both hysteresis losses and eddy-current losses in the steel. As we have seen, hysteresis losses depend on several factors including the frequency, the peak flux density, the type of core steel used, and the orientation of the flux with respect to the "grain" of the steel. All of the above factors, except the frequency, are under the control of the transformer designer. Core losses are sometimes referred to as *iron losses* and are commonly referred to as *no-load losses*, because core losses do not require load currents.

Decreasing the induced voltage per turn can reduce the peak flux density.

This obviously involves increasing the numbers of turns in both the primary and secondary windings in order to maintain the same transformer turns ratio. As we saw earlier, the disadvantage of adding more turns is that this increases the length of conductor and increases the conductor resistance. More cross-sectional area is required in order to keep the resistance constant. Doubling the number of turns requires about four times the volume of copper.

Another way of reducing core losses is to use various types of low-loss core steels that are now available, including "amorphous" core materials, which have extremely low losses and superior magnetic properties. Unfortunately, amorphous core materials have ceramic-like properties, so fabricating transformer cores with these materials is much more difficult than with laminated steel cores.

With grain-oriented steel, the direction of the core flux must be kept more or less parallel to the grain of the steel by mitering the corners of the laminations where the flux changes direction by 90°. Since the flux will cross the grain at about a 45° angle at the mitered edges, the hysteresis losses will increase somewhat in these places. These additional localized core losses must be factored into the calculation of the total core losses.

Building up the core with thin laminated strips controls eddy losses in the core, each strip having an oxide film applied to the surface. The oxide film is extremely thin and it is more like a high-resistance film than true electrical insulation; but since the potential differences between adjacent laminations is quite small, the resistance of the oxide film is very effective in breaking up the eddy current paths.

During the manufacture of the core, the core cutting machine must not be allowed to get dull; otherwise, "burrs" will form along the edges of the laminations. Burrs are imperfections that form electrical bridges between the laminations and create paths for eddy currents and increased losses. Sometimes the eddy currents near a burr can be large enough to cause localized overheating that can actually cause core damage.

Core losses are approximately proportional to the square of the excitation voltage E applied to the transformer. Therefore, placing an equivalent linear conductance G_m across the transformer terminals can approximate transformer core losses. The core losses are expressed by

$$W_m = E^2 G_m \tag{3.3.1}$$

Example 3.1

A certain large transformer core has a square cross section 24 in. on each side and is assumed to have a uniform flux density. At the normal operating voltage, the flux produces 90 V per turn (RMS). Each core lamination has an oxide film on one surface that is 1×10^{-4} in. thick and the total thickness of each lamination is 0.012 in. Determine the voltage difference between adjacent laminations along the perimeter of the core. Calculate the resistivity ρ of the oxide film required to limit the maximum circulating current losses to less than 1 mW/in. of surface along the perimeter of the core.

There are 2000 laminations in the core:

$$n = \frac{24 \text{ in.}}{0.012 \text{ in.}} = 2000$$

Since steel is a good electrical conductor, the voltage through the steel itself can be assumed to be nearly zero, so that nearly all of the voltage induced around the perimeter is developed across the layers of oxide film between the laminations. The voltage across each side of the lamination stack is 50% of the volts per turn induced around the perimeter of the core. Therefore, the voltage difference between each pair of laminations is calculated below. Refer to Figure 3.5.

$$\begin{aligned}\text{Volts per lamination} &= 50\% \times 90 \text{ V per turn}/2000 \text{ laminations}\\ &= 0.0225 \text{ V/lamination}\end{aligned}$$

We want to limit the losses W to 1 mW/in.^{2}:

$$W = \frac{E^2}{R} = \frac{(0.0225 \text{ V})^2}{R} < 1 \times 10^{-3} \text{ W/in.}^2$$

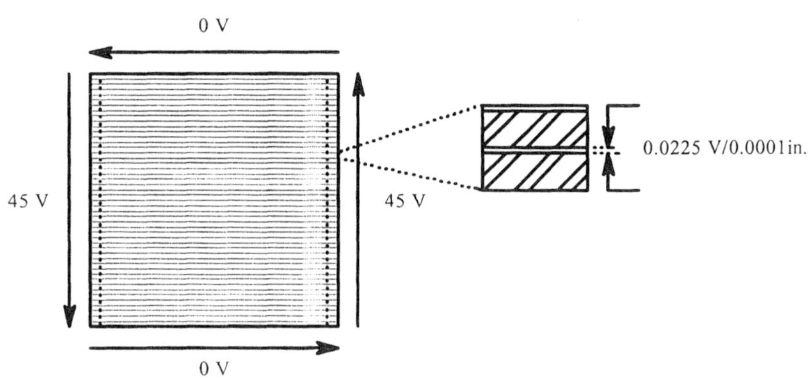

FIGURE 3.5 Core with 24 in.^{2} cross section used in Example 3.1.

Solving for R,

$$R > \frac{(0.0225\ \text{V})^2}{10^{-3}\ \text{W/in.}^2} = 0.506\ \Omega\text{-in.}^2$$

$$R = \rho \times \text{oxide thickness} = \rho \times 10^{-4}\ \text{in.}$$

$$\rho > 0.506\ \Omega\text{-in.}^2/10^{-4}\ \text{in.} = 5060\ \Omega\text{-in.}$$

For comparison, the resistivity of graphite is around 3500 Ω-in.; therefore, an oxide film with the minimum resistivity allowed would actually be a fairly good electrical conductor. It is a good idea to design the core with some electrical conductivity between the laminations to prevent static charges from building up between the laminations. Static charges in the core are allowed to bleed off to ground through an electrical connection at a single point, called the *core ground*.

3.4 MAGNETIZING REACTANCE

For an ideal transformer, the magnetizing current is assumed to be negligible. For a real transformer, some magnetizing current must flow when voltage is applied to the winding in order to establish a flux in the core. The voltage induced in the winding by the flux restrains the magnetizing current.

It was shown earlier that the magnetizing current is not really sinusoidal, but contains many odd harmonics in addition to the fundamental frequency. If we neglect the harmonics and concentrate on the fundamental frequency, the magnetizing current in the winding lags the applied voltage by 90°. In a two-winding transformer, this is equivalent to placing a reactance X_m, called the *magnetizing reactance*, in parallel with the transformer terminals.

The peak value of the magnetizing current is determined from the *B-H* curve of the core, which we have seen is very nonlinear. Therefore, the magnetizing reactance is not a constant but is voltage dependent; however, if the peak flux density is kept well below the point of saturation, X_m can be approximated by a constant reactance in most engineering calculations.

It is generally desirable to maximize X_m in order to minimize the magnetizing current. We saw earlier that inductance is inversely proportional to the reluctance of the core along the flux path and the reluctance of an air gap is several thousand times the reluctance of the same distance through the steel. Therefore, even tiny air gaps in the flux path can drastically increase the core's reluctance and decrease X_m. A proper core design must therefore eliminate all air gaps in the flux path. In Chapter 1 we saw how alternate layers of core steel are stacked so that flux is diverted around the gaps where laminations butt together. Since any flux that is diverted must flow between the laminations through their surfaces, it is vital that these surfaces lie perfectly flat against each other. All ripples or waves must be eliminated by compressing the core laminations together tightly. This also points out why the oxide layers on the lamination surfaces must be extremely thin: since these layers have essentially the same permeability as air and since the flux that is diverted from the air gaps must then travel through these oxide layers, the core's reluctance would greatly increase if these layers were not kept extremely thin.

3.5 EQUIVALENT CIRCUIT OF A TWO-WINDING TRANSFORMER

So far, the following elements of the equivalent circuit have been discussed in some detail: the leakage reactance, the load losses, the core losses and the magnetizing reactance. The leakage reactance is expressed as an inductive reactance in series with the secondary winding, but it is more correctly shown as split between the primary winding X_p and the secondary winding X_s. Similarly, the load losses are expressed as resistance R_p, in series with the primary winding, and resistance R_s, in series with the secondary winding. An ideal transformer is connected behind the series impedances. The magnetizing reactance X_m is approximated by a constant value, and the core losses are represented by a conductance G_m. These last two elements are shunted in parallel with the ideal transformer. The complete equivalent circuit of the transformer is shown in Figure 3.6.

This equivalent circuit can model all of the electrical properties of a two-winding transformer. This equivalent circuit also applies to all three-phase

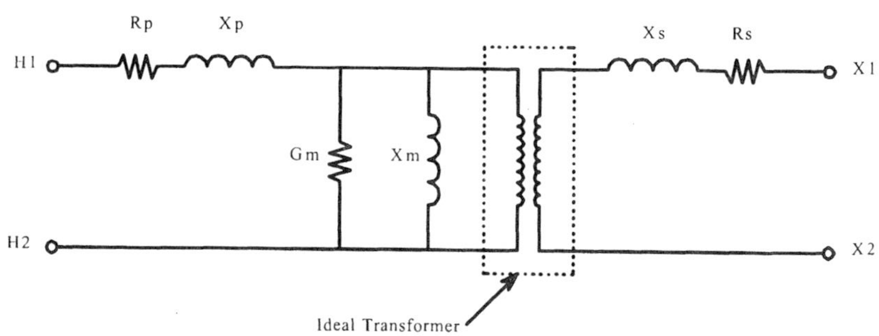

FIGURE 3.6 Equivalent circuit of a two-winding transformer in positive phase sequence and negative phase sequence networks.

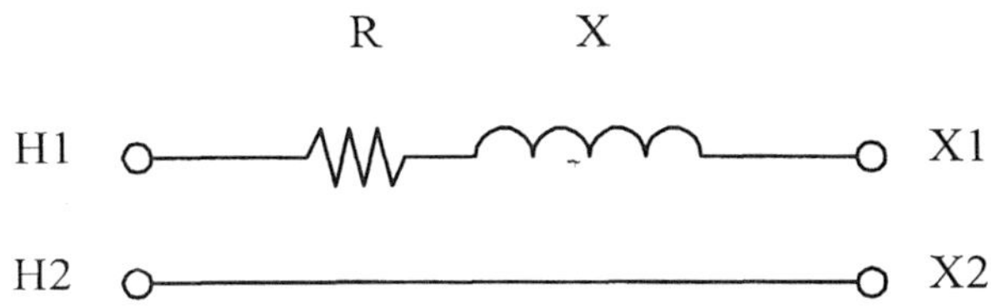

FIGURE 3.7 Simplified equivalent circuit of a two-winding transformer in positive phase sequence and negative phase sequence networks.

transformers on a per-phase basis. It must be mentioned here, however, that the equivalent circuit shown in Figure 3.6 is actually valid only in *positive phase sequence networks* and in *negative phase sequence networks*. A slightly different equivalent circuit must used in *zero phase sequence networks*. The meanings of these terms will be explained in Section 3.6.

Since typical values of magnetizing reactance are several orders of magnitude greater than typical values of leakage reactance, and since no-load losses are usually much smaller than load losses when the transformer is carrying rated load, the elements G_m and X_m can be neglected in most calculations. If voltages and currents are expressed in per unit instead of using actual system voltages and currents, the ideal transformer in the model will have a 1 : 1 turns ratio and can thus be eliminated from the equivalent circuit. Since R_p, R_s, X_p, and X_s are now all directly in series with each other, the resistance elements can be combined into a single resistance R and the reactance elements can be combined into a single reactance X. The result is a much-simplified equivalent circuit for the positive-sequence and negative-sequence networks, as shown in Figure 3.7.

The equivalent circuits of transformers modeled in zero phase sequence networks are different from those in positive and negative phase sequence networks. The zero-sequence equivalent circuit depends on the type of winding connections used; these equivalent circuits will be developed a little later in this chapter.

Before continuing with transformer equivalent circuits, it is necessary at this point to digress with a brief tutorial on symmetrical components.

3.6 A BRIEF TUTORIAL ON SYMMETRICAL COMPONENTS

The symmetrical component method is one of several methods that can be used to analyze unbalanced three-phase systems. The theory of symmetrical components is presented in many electrical engineering textbooks; unfortunately, there are few textbooks that present this topic in a practical and useful way that can easily be understood. Therefore, a brief tutorial on symmetrical components is presented here that will provide an understanding of positive, negative, and zero phase sequence voltages and currents and how the concept of symmetrical components is applied to transformers.

When three-phase circuits are balanced, calculations of voltages and currents are simple and straightforward. In these instances, the actual circuit and its *positive phase sequence network* are one and the same thing. The positive phase sequence network is nothing more than a network where the phase rotation is positive (normally *A-B-C*) and the phases are balanced; i.e., the voltages of all three phases have the same magnitudes, the currents of all three phases have the same magnitudes, and the phase angles' three phases are 120° apart.

Although the ideal set of balanced conditions may exist (approximately) under normal conditions, this is rarely the case for design-limiting conditions such as system faults. When voltages, currents, or impedances are unbalanced, the solutions become much more complex, especially when there is mutual induction between various parts of the circuit, as in the case of three-legged core form transformers with the three phases coupled magnetically through the core.

There have been several methods for attacking three-phase circuit problems for unbalanced conditions, but the most successful method in use today is called *symmetrical components*. In this method, an unbalanced circuit is represented by a combination of three balanced circuits, called *sequence networks*. The positive phase sequence network has already been described above. In the negative phase sequence network, the voltages and currents of all three phases are balanced but the phase rotation is opposite to the positive phase sequence; i.e., if the positive phase sequence is *A-B-C*, then the negative phase sequence is *C-B-A*. It is usually assumed that there are no negative or zero phase sequence voltage sources.

In the zero-phase sequence network, the voltages and currents of all three phases are identical and are all *in phase* with each other; i.e., there is no phase rotation. Since the phase currents are in phase, these add together at the neutral. Therefore, zero phase sequence networks always involve neutral or ground currents. Zero-sequence currents exist only if there are physical ground current paths in the circuit.

A set of unbalanced phase currents, I_a, I_b, and I_c, may be transformed into a set of symmetrical components I_0, I_1, and I_2, where I_0 is the zero phase sequence current, I_1 is the positive phase sequence current, and I_2 is the negative phase sequence current. The transformation is accomplished using the following set of equations:

$$I_0 = {}^1\!/\!_3(I_a + I_b + I_c) \tag{3.6.1}$$

$$I_1 = {}^1\!/\!_3(I_a + I_b\angle 120^\circ + I_c\angle 240^\circ) \tag{3.6.2}$$

$$I_2 = {}^1\!/\!_3(I_a + I_b\angle 240^\circ + I_c\angle 120^\circ) \tag{3.6.3}$$

The transformations shown in Eqs. (3.6.1) to (3.6.3) result in sequence currents that are referenced to *A* phase. There is also a set of sequence currents that is referenced to *B* phase, and a set of sequence currents referenced to *C* phase as well. The symmetrical component method does not require explicitly solving all three sets of sequence currents, because these sets of sequence currents are identical to each other except that they are shifted in phase. Therefore, it is implicitly understood that the sequence currents I_0, I_1, and I_2 are always referenced to *A* phase.

To transform a set of (*A* phase) sequence currents back into phase currents, the following equations are used:

$$I_a = I_0 + I_1 + I_2 \tag{3.6.4}$$

$$I_b = I_0 + I_1\angle 240^\circ + I_2\angle 120^\circ \tag{3.6.5}$$

$$I_c = I_0 + I_1\angle 120^\circ + I_2\angle 240^\circ \tag{3.6.6}$$

Voltages are transformed from phase values to sequence values and vice versa using exactly the same formulas, except that *E* quantities are substituted for *I* quantities in the equations. So far, it may seem that all this has accomplished is additional work; however, the additional work usually makes calculations involving unbalanced circuits much easier than dealing with phase quantities alone. We will see a direct application of this method at the end of this chapter.

The following steps summarize the technique of symmetrical components:

1. Construct the Thévenin equivalent circuits of the positive, negative, and zero phase sequence networks at the point of the fault.
2. Connect the Thévenin equivalent circuits constructed in step 1 together at the point of the fault, and calculate the sequence voltages and currents in all three networks at the point of the fault.
3. Transform the sequence voltages and currents at the point of the fault calculated in step 2 back into phase voltages and currents using Eqs. (3.6.4) to (3.6.6).

The Thévenin equivalent sequence networks are connected in different ways in step 2 in order to analyze various kinds of faults and unbalanced conditions. Three common unbalanced conditions are the single phase to

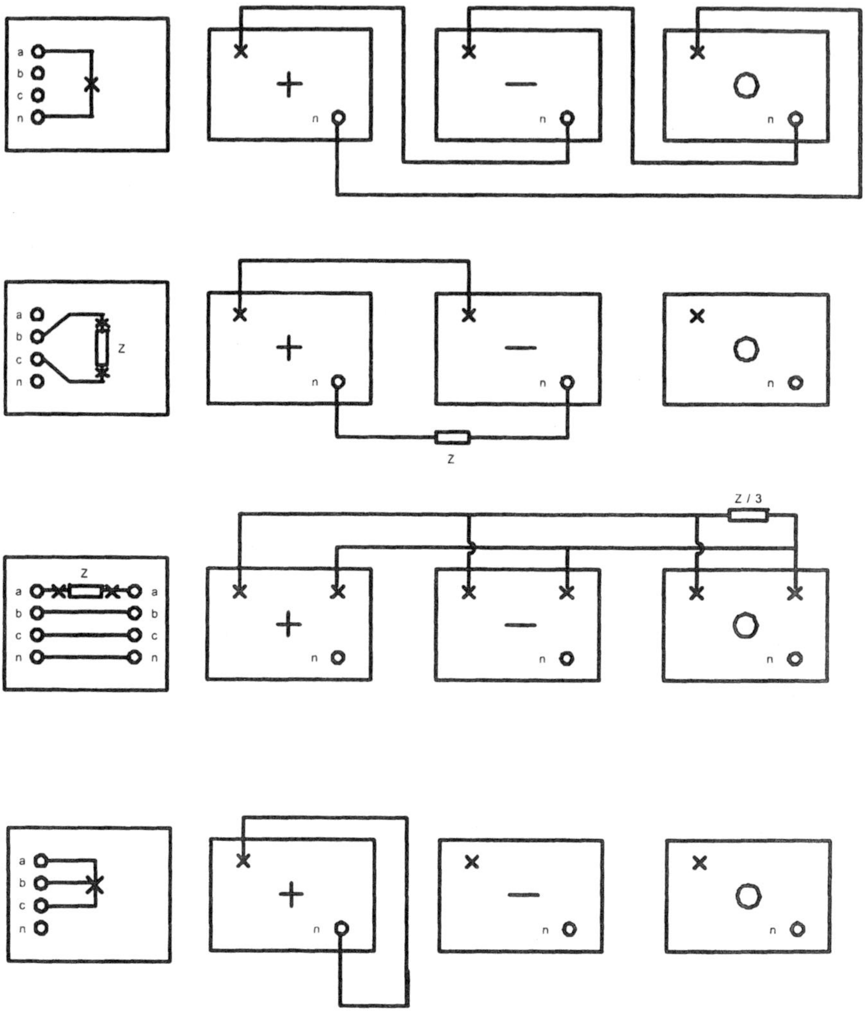

FIGURE 3.8 Sequence network connections for the three most common unbalanced conditions.

ground fault, a phase-to-phase connection through an impedance, and an impedance in series with one phase. The corresponding sequence network connections for these unbalanced conditions are shown in Figure 3.8.

The physical circuits are represented by the rectangles along the left of the figure. The positive, negative, and zero phase sequence networks are represented by the rectangles labeled +, −, and 0.

3.7 TRANSFORMER EQUIVALENT CIRCUITS MODELED IN ZERO PHASE SEQUENCE NETWORKS

Earlier in this chapter we developed the two-winding transformer equivalent circuit for the positive and negative phase sequence networks, as shown in Figure 3.7. This equivalent circuit is suitable for virtually any two-winding transformer, regardless of whether Y, Δ, or zigzag connections are used and regardless of the type of core construction used. We shall soon see that in zero sequence networks, the transformer equivalent circuits can be very different depending on the connections and type of core construction. This section will show how to develop the zero phase sequence equivalent circuits for the various types of transformers.

The basic equivalent circuit consists of series impedance elements between the primary and secondary windings, and a shunt impedance. The shunt impedance is approximately equal to the *open-circuit impedance* Z_{oc} measured at the primary windings with the secondary windings open-circuited, or measured at the secondary windings with the primary windings open-circuited. If per-unit values are used, Z_{oc} is approximately the same whether it is measured at the primary or secondary winding. The basic zero phase sequence transformer equivalent circuit is shown in Figure 3.9.

As we will soon see, the value of the shunt impedance is radically different for different types of core construction and winding connections.

If voltages are balanced and have an *A-B-C* or a *C-B-A* phase rotation, then Z_{oc} seen from the high-voltage side of the transformer is given by the following equation:

$$Z_{oc} = R_p + jX_p + \frac{1}{G_m - jX_m^{-1}} \tag{3.7.1}$$

The quantities G_m and X_m are the magnetizing conductance and reactance. In most cases G_m is practically zero and the magnitude of X_m is much greater than the magnitudes of R_p and X_p. Therefore, the open-circuit impedance of

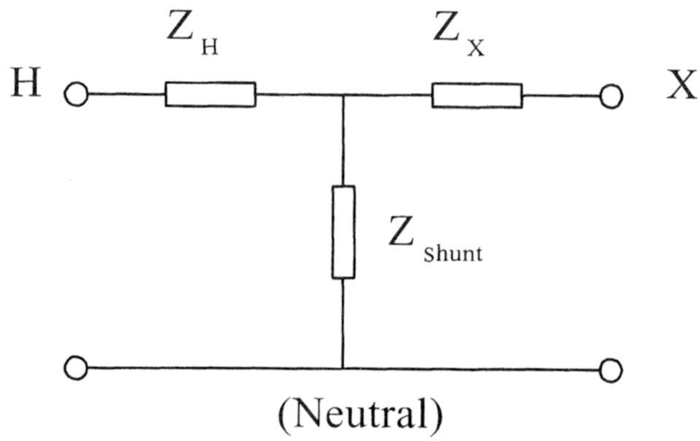

FIGURE 3.9 Equivalent circuit of a two-winding transformer in a zero phase sequence network.

a transformer in positive or negative phase sequence networks is approximately equal to the magnetizing impedance:

$$Z_{oc} \approx jX_m \tag{3.7.2}$$

In zero phase sequence networks, Z_{oc} is also approximately equal to jX_m for banks of three single-phase transformers; three-phase, five-legged core form transformers; and shell form transformers connected in a Y-Y.

For transformers connected Y-Δ and for three-phase, three-legged core form transformers connected Y-Y, the zero- sequence Z_{oc} is much smaller than the magnetizing impedance. In the case of a Y-Δ transformer bank, replacing the individual transformers with their equivalent circuits and applying a zero-sequence voltage to the primary connection easily shows this. The result is that there are three current paths of a similar nature: current flows from the H1 terminal through the transformer's series impedance elements, out the X1 terminal into the secondary delta connection, back through the X2 terminal of a transformer of a different phase, and out through that transformer's H2 terminal to ground. The arrows in Figure 3.10 show one of these current paths.

Therefore, the open-circuit zero-phase sequence impedance of a Y-Δ connected transformer on a per-phase basis is essentially equal to the series impedance of the transformer. The series impedance of a transformer is a small fraction (typically 5 to 10%) of the transformer base impedance. Recalling from Chapter 2 that a Y-Δ transformer is used as a grounding bank, it is now easy to understand how the grounding bank works. The Y-connected primary

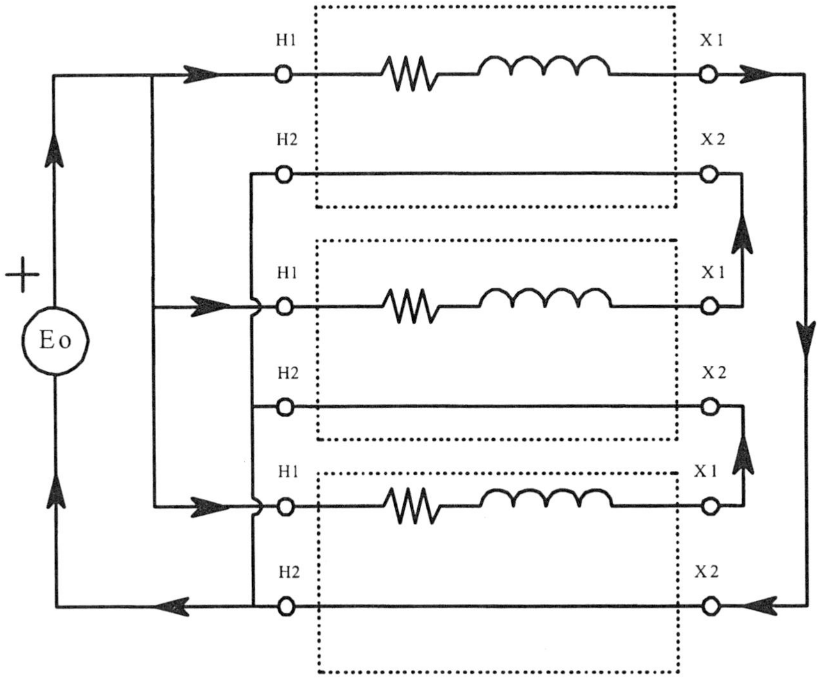

FIGURE 3.10 A Y-Δ transformer bank excited by a zero phase sequence voltage with the arrows showing the three current paths.

winding presents only a small impedance to zero phase sequence currents, which are the same as ground currents.

A similar result applies to the open-circuit impedance of a three-phase, three-legged core form transformer with the winding Y connected. When a zero phase sequence voltage E_0 is applied across all three windings, the magnetizing flux in the three phases arrive at the top core yoke in phase with each other without any physical magnetic return path to the bottom core yoke besides air. Since the path through air has a very high reluctance, the winding impedance (which is inversely proportional to the reluctance) is quite low, typically around 40 to 50% of the transformer's impedance base.

In the three-phase, five-legged core design, the flux generated by Y-connected windings excited by a zero phase sequence voltage E_0 will encounter a low-reluctance path through the outer core legs. Consequently, the open-circuit impedance of the Y-connected winding in a zero phase sequence network is about the same as the magnetizing impedance at low excitation levels.

A three-phase, three-legged core form transformer is shown in Figure 3.11 on the left. A five-legged core form transformer is shown in Figure 3.11 on the right. Zero phase sequence voltages excite both of these transformers. The zero phase sequence flux paths are shown as dotted lines.

In a five-legged core form transformer, the outer core legs and top and bottom core yokes do not usually have enough cross-sectional area to carry all of the flux of one phase. Since the zero-sequence flux of all three phases are in phase with each other, the core will go to into saturation if a zero-sequence exciting voltage is near the rated voltage.

This same effect exists for three-phase shell form transformers when excited by zero-sequence voltages. The lateral core legs between the phases are sized to carry the flux of one phase; however, the flux doubles when the zero-sequence fluxes from two phases arrive in phase at the lateral core legs. This could easily drive the core into saturation. Therefore, Z_{oc} for both five-legged core form and three-phase shell form transformers may decrease dramatically if excited by zero-sequence voltages near 100% of the rated voltage. The exact value of Z_{oc} is uncertain at best in these cases.

The series transformer impedance for zero phase sequence networks also depends on how the transformer is connected. For a Y-Y connection with the

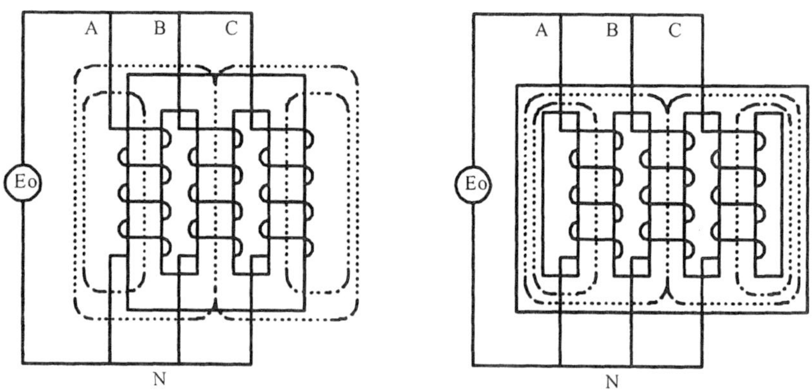

FIGURE 3.11 A three-legged core form transformer and a five-legged core form transformer excited by a zero phase sequence voltage showing the flux paths (dotted lines).

neutrals brought out, the series impedance in zero phase sequence networks is approximately the same as the series impedance in the positive phase sequence equivalent circuit. For Δ-Y and Y-Δ connections and for Y-Y connections with the neutrals not brought out, the series transformer impedance in the zero phase sequence equivalent circuit is infinite because there are no neutral paths for zero-sequence currents to flow. The diagrams in Figure 3.12 show the zero phase sequence equivalent circuits for various transformer connections.

$Z_H + Z_X$ is equal to the leakage reactance plus the primary and secondary winding resistances. Typical values for power transformers are between 7 and 15% of the transformer impedance base. As previously discussed, the magnetizing impedance Z_M varies depending on the core construction. For three-legged core form transformers, Z_M is typically between 40 and 50% of the transformer impedance base. For five-legged core form and three-phase shell form transformers, Z_M has very large values if the applied zero-sequence voltage is small, but Z_M decreases rapidly if the core goes into saturation as the applied zero-sequence voltage approaches 100% of rated voltage. Saturation occurs because the zero-sequence fluxes in the three core legs add in phase instead of canceling each other. This situation can occur when a Grd.Y- Grd.Y transformer is energized by a three-phase source with an open phase.

Example 3.2

There are five spare three-phase Y-Y low tension network (LTN) transformers and three-phase pad mount transformers that may be either three-legged core form or 5-legged core form designs. It is desired that the types of core construction of these transformers be determined prior to putting them into service. Using a single-phase 120 VAC 60 Hz source and an ammeter, devise a test to determine which types of core construction these transformers have.

The three high-voltage bushings (H1, H2, and H3) are jumpered together and connected to the "hot" leg of the 120 VAC 60 Hz source in series with the ammeter. The neutral bushing (H0) was connected to the grounded neutral leg of the 120 VAC source. The low-voltage bushings were open-circuited. The following current measurements were recorded for the five transformers.

Transformer 1
1500 KVA 12470Y-480Y
Kuhlman S/N 2–37930 Live-Front Pad
5.75% Impedance
I = 88 mA

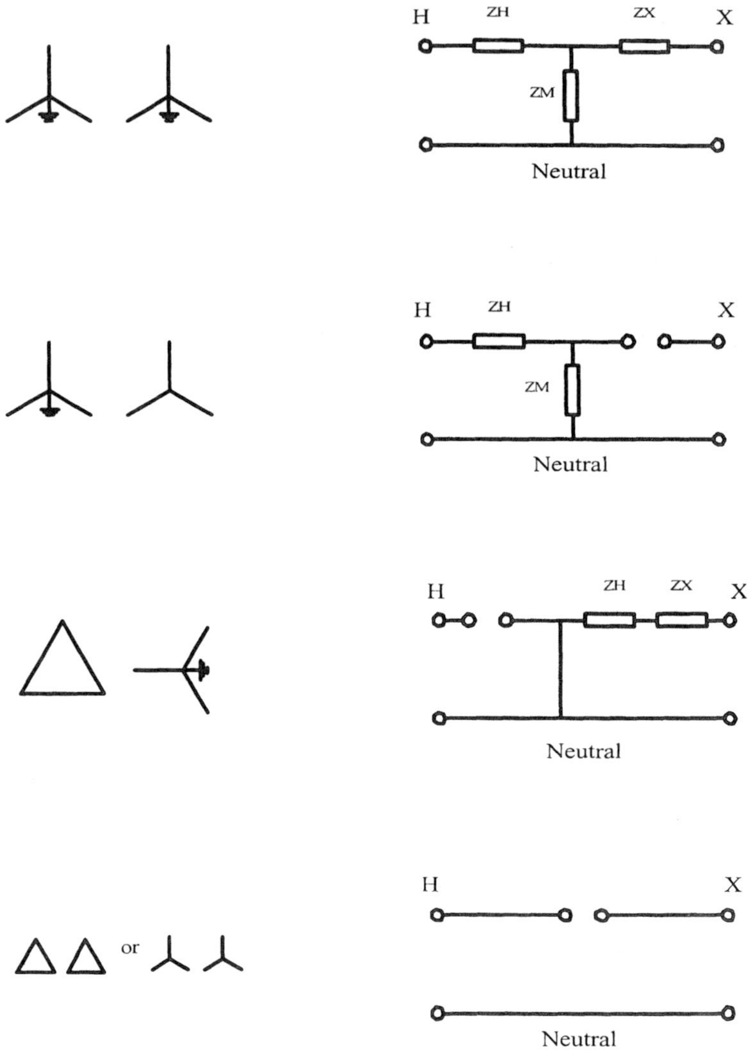

FIGURE 3.12 Zero phase sequence equivalent circuits for four common three-phase transformer connections.

Transformer 2
300 KVA 12470Y-208Y
Whse S/N 85JG73093 Dead-Front Pad
1.92% Impedance
I = 18 mA

Transformer 3
500 KVA 13000Y-216Y
G.E. S/N H323543P71AA LTN
4.83% Impedance
I = 21 mA

Transformer 4
300 KVA 12470Y-208Y
Allis Chalmers S/N 1729698 LTN
4.7% Impedance
I = 1.43 A

Transformer 5
G.E. S/N F645103-67P LTN
4.94% Impedance
I = 25 mA

Transformer 1 was verified to have a five-legged core design through a physical inspection of the core and coil assembly. Based on the measured currents, all the transformers have zero-sequence open-circuit currents less than 100 mA with the exception of transformer 4, which had a zero-sequence open-circuit current of 1.43 A. Based on the fact that its current is so much higher than the other transformers, it is concluded that transformer 4 is a three-legged core form transformer while the other transformers have the five legged core form design.

Example 3.3

Calculate the zero-sequence open-circuit impedance Z_{oc} of the three-legged core form transformer 4 and the five-legged core form transformer 1 in Example 3.2, expressing Z_{oc} as a per-unit impedance value of the transformer base impedance.

Transformer 4
Full-load current = 1.732 × 100,000 VA per phase/12,470 V = 13.89 A per phase

The measured open-circuit current is 1.43/3 A per phase = 0.477 A per phase.
$Z_{oc} = (1.732 \times 120 \text{ V}/12{,}470 \text{ V}) \times (13.89 \text{ A}/0.49 \text{ A}) = 0.472$ per unit

Transformer 1
Full-load current = 1.732 × 500,000 VA per phase/12,470 V = 69.45 A per phase
The measured open-circuit current is 0.088/3 A per phase = 0.0293 A per phase.
$Z_{oc} = (1.732 \times 120 \text{ V}/12{,}470 \text{ V}) \times (69.45 \text{ A}/0.0293 \text{ A}) = 39.5$ per unit

Referring to the zero-sequence equivalent circuit shown in Figure 3.9, Z_{oc} is equal to $Z_H + Z_{shunt}$. It is easily seen in this example how the zero-sequence equivalent circuits are affected by the core design.

3.8 SERIES IMPEDANCE AND REGULATION

The series impedance of a transformer consists of a resistance that accounts for the load losses and a reactance that represents the leakage reactance. This impedance has a very low power factor, consisting almost entirely of leakage reactance with only a small resistance. As discussed earlier, the transformer design engineer can control the leakage reactance by varying the spacing between the windings. Increasing the spacing "decouples" the windings and allows more leakage flux to circulate between the windings, increasing the leakage reactance.

While leakage reactance can be considered a transformer loss because it consumes reactive power, some leakage reactance is necessary to limit fault currents. On the other hand, excessive leakage reactance can cause problems with *regulation*. Regulation is often defined as the drop in secondary voltage when a load is applied, but regulation is more correctly defined as the *increase* in secondary output voltage when the load is *removed*. The reason that regulation is defined this way is that transformers are considered to be "fully loaded" when the secondary output voltage is at the rated secondary voltage. This requires the primary voltage to be greater than the rated primary voltage at full load.

Let E_p equal the primary voltage and let E_s equal the secondary voltage when the transformer is fully loaded. Using per-unit values instead of primary and secondary voltage values, the per-unit secondary voltage will equal E_p with the load removed. Therefore, the definition of regulation can be expressed by the following equations.

$$\text{Regulation} = \frac{E_p - E_s}{E_s} \tag{3.8.1}$$

Since $E_s = 1$ by definition,

$$\text{Regulation} = E_p - 1 \tag{3.8.2}$$

Regulation depends on the power factor of the load. For a near-unity power factor, the regulation is much smaller than the regulation for an inductive load with a small lagging power factor.

Example 3.4

A three-phase 1500 KVA 12470Y-208Y transformer has a 4.7% impedance. Calculate the three-phase fault current at the secondary output with the primary connected to a 12,470 V infinite bus. Calculate the regulation for a power factor of 90% at full load.

The three-phase fault is a balanced fault, so the positive-sequence equivalent circuit applies. The full-load secondary current is calculated as follows:

$$I = 1.732 \times 500{,}000 \text{ VA per phase}/208 \text{ V} = 4167 \text{ A per phase}$$

The per-unit fault current is the primary voltage divided by the series impedance:

$$\frac{1}{0.047} = 21.27 \text{ per unit}$$

The secondary fault current is equal to the per-unit fault current times the full-load current:

$$I_f = 21.27 \text{ per unit} \times 4167 \text{ A per phase} = 88{,}632 \text{ A per phase}$$

To calculate regulation, the secondary voltage is $1\angle 0°$ per unit by definition. Applying a 1 per unit load at a 90% lagging power factor, $I = 1.0\angle -25.8°$. Since the series impedance is mainly inductive, the primary voltage at full load E_p can be calculated as follows:

$$\begin{aligned} E_p &= 1\angle 0° + 1.0\angle -25.8° \times 0.047\angle 90° \\ &= 1.02 + j0.042 = 1.021 \text{ per unit} \end{aligned}$$

$$\text{Regulation} = E_p - 1 = 0.021 = 2.1\%$$

3.9 MATCHING TRANSFORMERS FOR PARALLEL AND BANK OPERATIONS

The following rules must be obeyed in order to successfully connect two or more transformers in parallel with each other:

1. The turns ratios of all of the transformers must be nearly equal.
2. The phase angle displacements of all of the transformers must be identical.
3. The series impedances of all transformers must be nearly equal, when expressed as "%Z" using the transformer impedance base.

The first two rules are required so that the open-circuit secondary voltages of the transformers are closely matched in order to avoid excessive circulating currents when the parallel connections are made.

The last rule is based on the fact that for a given voltage rating and %Z, the ohmic impedance of a transformer is inversely proportional to its KVA rating. When transformers having the same %Z are connected in parallel, the load currents will split in proportion to the KVA ratings of the units. Therefore, transformers with different KVA ratings can be successfully operated in parallel as long as their %Z values are all approximately the same.

Example 3.5

(*This example is based on an actual event.*)

Two three-phase 10,000 KVA 66,000Δ-12,470Y volt transformers were in parallel operation in a substation. The primaries of the two transformers are connected to a 66 kV transmission line through a single air break switch. This switch is designed to interrupt magnetizing current only, which is less than 1 A. The transformers were being removed from service and the secondary loads had been removed. A switchman then started to open the air break switch, expecting to see a small arc as the magnetizing current was interrupted. Instead, there was a loud "bang" and there was a ball of flame where the air break switch contacts had vaporized. Something was obviously wrong.

Upon closer inspection, it was revealed that the two transformers had been set on widely different taps: The first transformer was on the 62,700 V primary tap and the second transformer was on the 69,300 V primary tap. Both transformers had a 7% impedance. Because the turns ratios were unequal, a circulating current was set up even without any secondary load. The open-circuit secondary voltage difference, assuming 66 kV at the transformer primaries, is calculated below.

$$\Delta E_s = 66{,}000 \times \left(\frac{12{,}470}{62{,}700} - \frac{12{,}470}{69{,}300} \right) \text{V} = 1250 \text{ V} = 0.10 \text{ per unit}$$

The per-unit circulating current in the secondary loop is equal to ΔE_s divided by the sum of the per-unit impedances of the two transformers:

$$I_c = \frac{0.10}{0.14} = 0.714 \text{ per unit}$$

Converting I_c into amperes:

$$I_c = 0.714 \times 10{,}000 \text{ KVA}/(12.47 \text{ kV} \times 1.732) = 331 \text{ A per phase}$$

Since I_c flows in a loop in the secondary circuit, the current out of the secondary of the first transformer equals the current into the secondary of the second transformer. But since the turns ratios are not equal, I_c does not get transformed into equal and opposite currents at the primaries.

$$\text{Primary current of first transformer} = 331 \text{ A} \times \frac{12{,}470}{62{,}700} = 65.8 \text{ A per phase}$$

$$\text{Primary current of second transformer} = 331 \text{ A} \times \frac{12{,}470}{69{,}300} = 59.6 \text{ A per phase}$$

The net current through the air break switch, I_{AB}, is the difference in the primary currents:

$$I_{AB} = 65.8 \text{ A per phase} - 59.6 \text{ A per phase} = 6.2 \text{ A per phase}$$

The current through the air break switch supplies the $I_c^2X_s$ reactive losses of both transformers and therefore lags the primary voltage by 90°. The resulting current exceeded the interrupting rating of the switch, causing it to fail. The conditions described in this example are diagramed in Figure 3.13.

3.10 IMPEDANCE MISMATCH IN THREE-PHASE TRANSFORMER BANKS

When connecting single-phase transformers in three phase banks, proper impedance matching becomes even more critical. In addition to following the three rules for parallel operation, it is also a good practice to try to match the *X/R* ratios of the three series impedances to keep the three-phase output voltages balanced.

If single-phase transformers are connected in a Y-Y bank with an isolated neutral, then the magnetizing impedances should also be equal *on an ohmic basis*. Otherwise, the transformer having the largest magnetizing impedance will have a highest percentage of exciting voltage, increasing the core losses of that transformer and possibly driving its core into saturation.

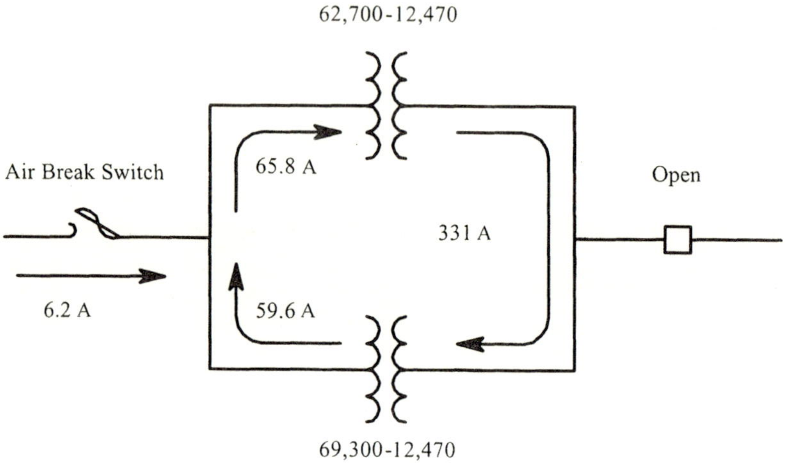

FIGURE 3.13 Circuit diagram for Example 3.5 showing circulating currents for parallel transformers with different turns ratios.

When single-phase transformers with the same KVA ratings are connected in a Y-Δ bank, impedance mismatches can cause a significant load unbalance among the transformers. The general case where all three transformers have different impedances is derived below.

Define the following ratios of impedances, w and x, as follows. (*Note*: In general, these ratios are complex numbers because the impedances are complex numbers.)

$$w \equiv \frac{Z_b}{Z_a} \quad \rightarrow \quad Z_b = wZ_a \tag{3.10.1}$$

$$x \equiv \frac{Z_c}{Z_a} \quad \rightarrow \quad Z_c = xZ_a \tag{3.10.2}$$

With a balanced three-phase load of 1 per unit applied, the phase currents are as follows (see Figure 3.14):

$$i_1 = 1\angle 0° \tag{3.10.3}$$

$$i_2 = 1\angle 240° \tag{3.10.4}$$

$$i_3 = 1\angle 120° \tag{3.10.5}$$

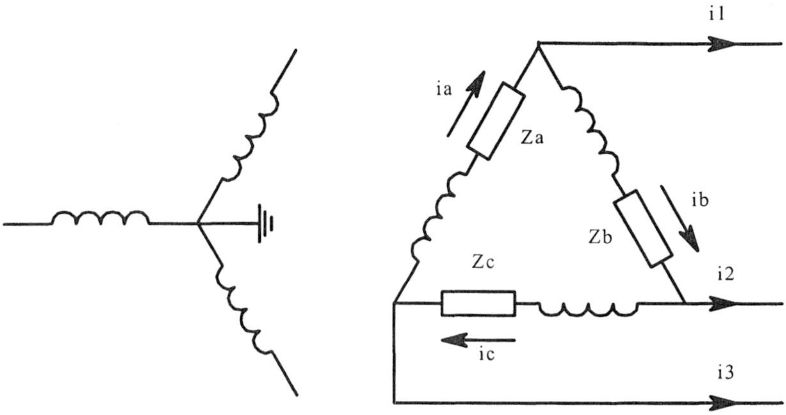

FIGURE 3.14 Y-Δ transformer bank made up with three single-phase transformers having different impedances.

The voltage drop around the closed Δ path must equal zero:

$$i_aZ_a + i_bZ_b + i_cZ_c = i_aZ_a + i_bwZ_a + i_cxZ_a = 0 \tag{3.10.6}$$

$$i_a + w\, i_b + x\, i_c = 0 \tag{3.10.7}$$

The load currents can be expressed in terms of the transformer currents i_a, i_b, and i_c:

$$i_a - i_b = i_1 = 1\angle 0° \tag{3.10.8}$$

$$i_c - i_a = i_3 = 1\angle 120° \tag{3.10.9}$$

The transformer secondary currents are solved from the last three equations:

$$i_a = \frac{w\angle 0° - x\angle 120°}{1 + w + x} \tag{3.10.10}$$

$$i_b = \frac{x\angle 240° - 1\angle 0°}{1 + w + x} \tag{3.10.11}$$

$$i_c = \frac{1\angle 120° - w\angle 240°}{1 + w + x} \tag{3.10.12}$$

The bank is balanced if $w = 1$ and $x = 1$. Even when the impedance magnitudes are all equal, if the *X/R* ratios are unequal then $w \neq 1$ or $x \neq 1$ and the transformer currents will not be equal. The most heavily loaded transformer

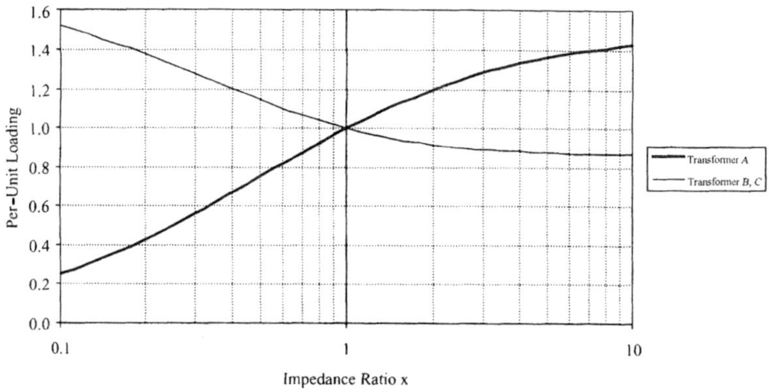

FIGURE 3.15 Chart showing the loading of a Y-Δ transformer bank made up with three single-phase transformers as a function of the impedance imbalance between two of the phases and the third phase.

limits the maximum load that can be applied to the bank, so any unbalance will reduce the load rating of the bank.

Let us examine the specific case where the *X/R* ratios of all three transformers are equal, and transformer *B* is identical to transformer *C* but the impedance of transformer *A* is not equal to the impedances of the other two transformers. In this case $Z_b/Z_a = Z_c/Z_a = x$ and x is a real number. The currents i_a, i_b, and i_c are multiplied by 1.732 to obtain the per-unit loading due to the Y-Δ connection. The graph in Figure 3. 15 plots the per-unit transformer loadings as x varies over the range 0.1 to 10.

3.11 TEMPERATURE RISE AND THE THERMAL CAPABILITY

Transformer KVA ratings have been alluded to on a number of occasions up to this point without explaining how the KVA rating is determined. The KVA rating of a transformer is simply the steady-state KVA load applied to the *output* of the transformer at the voltage rating of the output winding that produces an average winding temperature rise (above the ambient temperature) equal to 65°C. For older transformers, the rated average winding temperature rise was 55°C. Advances in insulating materials allowed a 10°C increase in average temperature. The temperature rise of the winding is caused by all of the transformer losses that were previously discussed in this chapter. Therefore, the winding temperature is a function of load losses and no-load losses.

The *thermal capability* of a transformer is defined in a slightly different way from the rated KVA. Thermal capability is the KVA load applied to the output of a transformer that causes the hottest area in the windings, called the *winding hot spot*, to reach some limiting temperature. The hot-spot temperature determines the rate of loss of life of the transformer as a whole, which is a cumulative effect. Therefore, the hot-spot temperature limit is usually based on a loss-of-life criterion. The standard method for calculating the thermal capability of oil-insulated transformers will be covered fully at the end of this chapter.

3.12 INTERPRETING TRANSFORMER TEST REPORTS

Much of the information needed to calculate temperature rise and the thermal capability of a transformer is contained in the *transformer test report*. When one purchase order is used to procure a number of identical transformers from the same manufacturer, some testing is performed on each transformer while certain other tests are performed only on one representative unit in that purchase order. There is a recommended format of a transformer test report [1]. Some transformer manufacturers follow the recommended format exactly while others prefer to use their own format, but the information contained in the test reports is standard throughout the industry.

A sample test report is shown in Figure 3.16. It was based on an actual factory test report for an order of four transformers used in a generating station. A certified test engineer would sign the actual transformer test report, testifying to its authenticity. In the upper portion of the test report is the general design information, such as the cooling class, number of phases, frequency rating, rated temperature rise, voltage ratings of the windings, and the KVA rating. The transformers shown in Figure 3.16 have an *FOA Cooling Class*, which is an abbreviation for forced-oil-air. This means that heat is transferred from the core and coils to the insulating oil, which is forced through the transformer by oil pumps. The heat is then transferred from the insulating oil to the ambient air through a set of radiators which have air forced through them by fans.

The voltage rating of the HV winding is 512500Grd.Y/296000, which means the primary winding is designed to be Y connected to a system having a phase-to-phase voltage of 512.5 kV and a phase-to-neutral voltage of 296 kV. The "Grd." designation usually indicates that the winding has reduced insulation near the neutral end, the neutral being normally grounded. The full winding rating from phase to ground is 296,000 V. Additional HV winding taps are available at 538,125, 525,312, 499,688, and 486,875 V.

The voltage rating of the LV winding is 23600, which means that it is

Acme Transformer Company

Power Systems Division

P.O. Box 123, Macungie, PA 18062

TRANSFORMER TEST REPORT

Customer GenCo Unit 2

Date of Test 8/25/78 Customer's Order 8-12345-1 Our Order A-6789-5

Type FOA Phase 1 Cycles 60 Rise 65°C Spec. 13140

H.V. Volts 512500Grd.Y/296000 L.V. Volts 23600 T.V. Volts

Taps 538125/525312/499688/486875 Taps Taps

KVA 405000 KVA 405000 KVA

Serial Number C-05213-5	**-1**	**-2**	**-3**	**-4**	**Avg.**	**Guarantee**
Polarity: Subtractive						
W.M. Copper Loss @ Full Load 85°C	909932	905000	910760	898550	906060	
Core Loss @ 100% Voltage	280488	28466	285360	299280	287448	295000
Total Loss @ Full Load & 100% Volts.	1190420	1189664	1196120	1197830	1193508	1235000
Core Loss @ 110% Voltage	427344	431520	427344	451008	434304	
% Exciting current @ 100% Voltage	0.73	0.67	0.61	0.80	0.70	1.20
% Exciting current @ 110% Voltage	1.79	1.69	1.54	1.86	1.72	2.70
% Impedance @ 85°C	14.18	14.16	14.12	14.21	14.17	14.00
% Resistance @ 85°C	0.225	0.223	0.225	0.222	0.224	
% Reactance @ 85°C	14.18	14.16	14.12	14.21	14.17	
% Regulation @ 100% P.F. Full Load					1.23	1.30
% Regulation @ 80% P.F. Full Load					9.31	9.30
% Efficiency @ Full Load 100% P.F.					99.70	99.69
% Efficiency @ 3/4 Load 100% P.F.					99.73	99.72
% Efficiency @ 1/2 Load 100% P.F.					99.74	99.73
% Efficiency @ 1/4 Load 100% P.F.					99.66	
Insulation Test HV – LV/Core (for 1 min)	34000V	34000V	34000V	34000V		
Insulation Test LV – Core (for 1 min)	50000V	50000V	50000V	50000V		
HV Temp Rise Corrected to Shutdown °C				59.4		
LV Temp Rise Corrected to Shutdown °C				50.5		
Oil Rise °C				39.0		
Hot Spot Temperature Rise °C – HV				67.1		
Hot Spot Temperature Rise °C – LV				56.1		

FIGURE 3.16 Sample test report for four transformers on a single purchase order.

designed to be Δ-connected to a system having a phase-to-phase voltage of 23.6 kV. The full winding rating is the same as the phase-to-phase system voltage, 23,600 V.

We can glean a lot of additional information from this test report:

- The core losses represent about 24% of the total losses.
- The core losses increase by over 50% when the applied voltage increases by 10% over the rated voltage.
- The resistance portion is only about 1.6% of the total series impedance, so the series impedance can be approximated by the leakage reactance.
- The transformers are over 99% efficient at full load.
- All of the electrical characteristics of the four transformers are remarkably similar.

The methods used in performing factory tests will be covered in more detail in Chapter 8. "W.M. Copper Loss" is an abbreviation for watt meter copper loss. Since typical magnetizing current power factors are extremely small, core loss measurements require carefully calibrated test equipment.

At the bottom of the test report are the results of a *heat run test*. The heat run test applies a load equal to the nameplate KVA rating of the transformer at full voltage using an inductive load. Using an inductive load reduces the energy use during this test. Nonetheless, the energy requirements for supplying the transformer losses alone can be significant for large transformers. It is an expensive and time-consuming test, so it is usually only performed on one of the transformers on a purchase order. In the case of this sample test report in Figure 3.16, the heat run was only performed on Serial Number C-05213-5-4 and the results can then be used to calculate the thermal capabilities of the other transformers on this purchase order.

It was intended in the original design to build a transformer having a 65°C average winding temperature rise when continuously loaded at the 105,000 KVA nameplate rating. Note from the test report, however, that the HV winding only attained a 59.4°C rise and the LV winding only attained a 50.5°C rise. Therefore, the as-built transformer has a 5.6°C temperature design margin that can be used to increase the loading above the nameplate rating. It is this increased loading, and not the nameplate rating, that is the true *thermal capability* of that transformer. The average winding temperatures are not directly measured in this test. Instead, the winding temperatures are inferred from conductor resistance measurements taken immediately after the test is shut down. The dc resistance of a metallic conductor is a linear function of the conductor's absolute temperature. The temperature of a transformer winding is not uniform along the length of the conductor; however, the resistance measurement of the total conductor will provide the average temperature of the winding.

The last two entries on the test report are the hot spot temperature rise of the HV and LV windings. These values are not actually measured by any test, but are calculated based on the average winding temperature and the oil temperature. There is only about a 6.5°C temperature gradient between the hot spot temperature and the average winding temperature. This is fairly typical of transformers in the FOA cooling class. Since cooling oil is forced through the windings by pumps, the oil velocity is quite high. The high oil velocity tends to equalize the oil temperatures along the flow paths through the windings. For transformers without oil pumps, such as the self-cooled (OA class) or forced-air cooled only (FA class), the oil velocities are much slower and there is a much greater temperature gradient between the hot spot temperature and the average winding temperature.

3.13 CALCULATING THE HOT-SPOT TEMPERATURE USING THE IEEE METHOD

The following series of equations outlines the algorithm used to compute temperatures in oil-filled transformers [2]. This method can be applied to transformers of any MVA size. The ambient temperature is assumed to be constant and all temperatures shown are in °C. The hot-spot temperature θ_H is the temperature that determines the thermal capability of the transformer and is given by the following equation.

$$\theta_H = \theta_A + \Delta\theta_{\mathrm{TO}} + \Delta\theta_H \tag{3.13.1}$$

where

θ_A = ambient temperature
$\Delta\theta_{\mathrm{TO}}$ = temperature gradient of the top oil temperature over the ambient temperature
$\Delta\theta_H$ = temperature gradient of the hot spot over the top oil temperature

The top oil temperature θ_{TO} is calculated as follows:

$$\theta_{\mathrm{TO}} = \theta_A + \Delta\theta_{\mathrm{TO}} \tag{3.13.2}$$

If the load changes from one value to another,

$$\Delta\theta_{\mathrm{TO}} = (\Delta\theta_{\mathrm{TO},U} - \Delta\theta_{\mathrm{TO},i})(1 - e^{-t/\tau_{\mathrm{TO}}}) + \Delta\theta_{\mathrm{TO},i} \tag{3.13.3}$$

where

$\Delta\theta_{\mathrm{TO},U}$ = ultimate steady-state top oil temperature gradient
$\Delta\theta_{\mathrm{TO},i}$ = initial top oil temperature gradient
t = duration of the load change, h
τ_{TO} = oil time constant, h

$$\Delta\theta_{\mathrm{TO},i} = \Delta\theta_{\mathrm{TO},R}\left(\frac{K_i^2 R + 1}{R + 1}\right)^n \tag{3.13.4}$$

where

R = ratio of load losses at rated load to no-load losses from the test report
K_i = ratio of the initial load to the rated load
$\Delta\theta_{\mathrm{TO},R}$ = top oil temperature gradient at rated load from the test report
n = an empirical exponent that depends on the cooling class

A generally accepted value for n is 0.8 for OA cooling class transformers, 0.9 for the FA and FOA cooling classes, and 1.0 for directed flow FOA or FOW (forced oil and water) cooling classes.

$$\Delta\theta_{\mathrm{TO},U} = \Delta\theta_{\mathrm{TO},R}\left(\frac{K_U^2 R + 1}{R + 1}\right)^n \tag{3.13.5}$$

where K_U = the ratio of the ultimate (final) load to the rated load.

The winding hot-spot temperature gradient is also load and time dependent:

$$\Delta\theta_H = (\Delta\theta_{H,U} - \Delta\theta_{H,i})(1 - e^{-t/\tau_W}) + \Delta\theta_{H,I} \tag{3.13.6}$$

where

$\Delta\theta_{H,U}$ = ultimate steady-state hot-spot temperature gradient
$\Delta\theta_{H,i}$ = initial hot spot temperature gradient
t = duration of the load change, h
τ_W = winding time constant, h

$$\Delta\theta_{H,i} = \Delta\theta_{H,R} \times K_i^{2m} \tag{3.13.7}$$

where

$\Delta\theta_{H,R}$ = the hot-spot temperature gradient at rated load from the test report
m = an empirical exponent depending on the cooling class

A generally accepted value for m is 0.8 for the OA, FA, and FOA cooling class transformers and 1.0 for the directed flow FOA or FOW (forced oil and water) cooling classes.

The oil time constant τ_{TO} is calculated as follows: Let C = the thermal capacity of the transformer. For OA and FA cooling classes,

$$C = 0.06 \times W_{CC} + 0.04 \times W_{TF} + 1.33 \times \text{GO Wh/°C} \tag{3.13.8}$$

and for the FOA cooling class,

$$C = 0.06 \times W_{CC} + 0.06 \times W_{TF} + 1.93 \times \text{GO Wh/°C} \tag{3.13.9}$$

where

W_{CC} = weight of the core and coil assembly, lb
W_{TF} = weight of the tank and fittings, lb
GO = total gallons of oil that cools the transformer

$$\tau_{TO,R} = C \times \frac{\Delta\theta_{TO,R}}{P_{T,R}} \tag{3.13.10}$$

where $P_{T,R}$ = total loss at rated load from the test report.

$$\tau_{TO} = \tau_{TO,R} \times \frac{\Delta\theta_{TO,U}/\Delta\theta_{TO,R} - \Delta\theta_{TO,i}/\Delta\theta_{TO,R}}{(\Delta\theta_{TO,U}/\Delta\theta_{TO,R})^{1/n} - (\Delta\theta_{TO,i}/\Delta\theta_{TO,R})^{1/n}} \tag{3.13.11}$$

The calculation of the winding time constant τ_W is not described in the IEEE standard. Most engineers consider the winding time constant to be very short, and a value of zero is used by some. Although the IEEE method seems rather complicated, it is straightforward enough with the use of a desktop computer. The IEEE standard includes a computer program in Basic code that performs the temperature calculations.

For a load that cycles periodically, the following method is used to calculate temperatures:

- The load is approximated by a series of step changes. Satisfactory results can be obtained with just a few step changes over the course of 24 h, but better accuracy will be achieved if the load cycle is broken into 1 h intervals.
- A set of initial temperatures is assumed for the beginning of the load cycle and the temperatures at the end of the first load step are calculated using Eqs. (3.13.1) to (3.13.7).
- This process is repeated for each of the load steps until the end of the cycle is reached.

- This process is repeated for each of the load steps until the end of the cycle is reached.
- The process is then repeated for the entire load cycle, using the temperatures calculated at the end of each load cycle as the initial temperatures of the next load cycle.
- Eventually, after perhaps two or three cycles, the temperatures at the end of the load cycle converge to the same values at the end of every cycle.
- When this state is reached, the solution has converged.

3.14 CALCULATING THE LOSS OF LIFE

Industry standards also address the loss of life of a transformer due to temperature and aging. The relation of insulation aging to time and temperature follows the well-known Arrhenius chemical reaction rate model.

The adaptation used in the IEEE standard [3] has the following form.

$$\text{Per unit life} = Ae^{B/(\theta_H+273)} \tag{3.14.1}$$

where θ_H = winding hot-spot temperature in °C, and A and B are constants.

The meaning of per unit life is illustrated as follows: If per unit life = 2, then the transformer would be expected to last twice the "normal" life. If per unit life = 0.5, then the transformer would be expected to last only half the "normal" life. Normal life for most transformers is considered to be around 30 to 40 years.

The constants A and B depend on the types of material used to insulate the windings. Since cellulose in the form of kraft paper is the most common insulation material, these constants have been worked out empirically:

$$\text{Per unit life} = 9.8 \times 10^{-18} e^{15000/(\theta_H+273)} \tag{3.14.2}$$

The winding hot-spot design temperature to attain a normal life is 110°C. This is based on an assumed ambient temperature of 30° plus the 65°C average winding temperature gradient over ambient plus a 15°C allowance for the hot-spot gradient over the average winding temperature. Using θ_H = 110°C yields a per unit life = 1.

The *aging acceleration factor* F_{AA} is the ratio of the per unit life at the design temperature of 110°C divided by the per unit life at some operating temperature θ_H.

The constant A then divides out:

$$F_{AA} = e^{15000/383-15000/(\theta_H+273)} = e^{39.16-15000/(\theta_H+273)} \tag{3.14.3}$$

To calculate the equivalent aging of the transformer F_{EQA} with a varying hot-spot temperature such as occurs for a cycling load or a seasonal ambient temperature, F_{AA} is integrated over time and the integral is divided by the total time to obtain the average.

The per unit life and F_{AA} are plotted vs. hot-spot temperature in the chart shown in Figure 3.17.

It should be stressed that most transformer failures are random events that occur for various reasons besides insulation loss of life. Therefore, the formula for per unit life cannot be used as a predictive model to determine when a given transformer will ultimately fail. However, it is indeed certain that overloading a transformer will shorten its life, so it is a good practice from

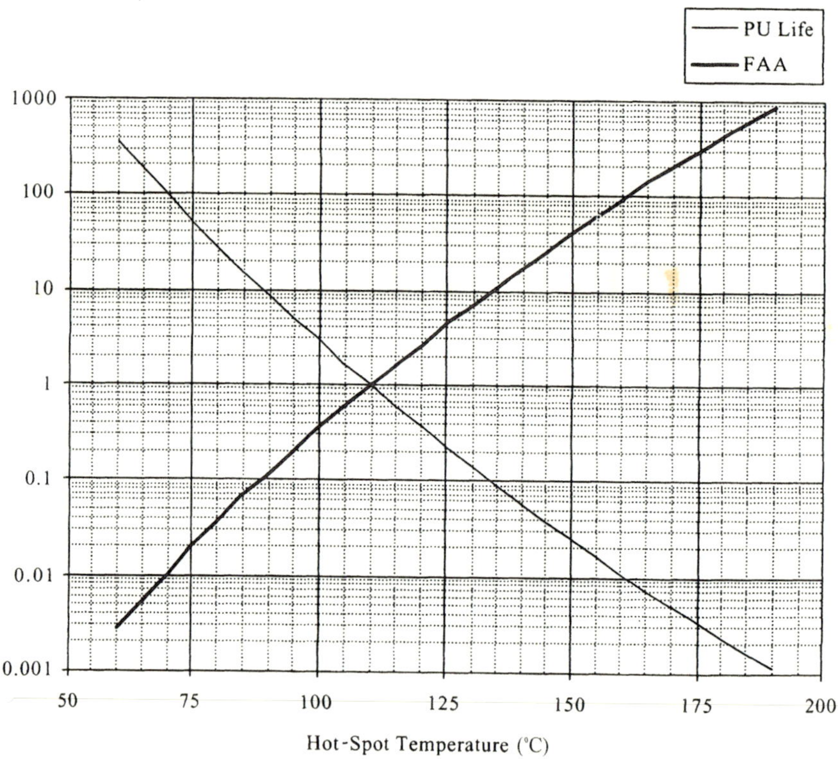

FIGURE 3.17 The per unit life and the aging acceleration factor as a function of the hot-spot temperature.

a reliability standpoint to keep the loading within the transformer's thermal capability.

Example 3.6

Using the transformer test report shown in Figure 3.16, calculate the hot-spot temperature of transformer S/N C-05213–5-1 at 2 h after the load is changed from 350 MVA to 450 MVA, assuming the initial load was applied for a long time prior to the change. Use the IEEE temperature calculation method assuming a constant 30°C ambient temperature. The following additional data apply to these transformers: weight of core and coils (W_{CC}) = 346,000 lb; weight of tank and fittings (W_{TF}) = 64,500 lb; gallons of oil (GO) = 13,400. Assume the cooling class is FOA with directed flow.

From the test report, it is clear that the HV winding has the higher temperature and is the limiting winding. The initial and ultimate temperatures are calculated first:

$$R = \frac{909{,}932 \text{ W}}{280{,}488 \text{ W}} = 3.244 \text{ from the test report}$$

$$K_i = \frac{350 \text{ MVA}}{405 \text{ MVA}} = 0.864$$

$$\Delta\theta_{\text{TO},R} = 39.0°\text{C from the test report}$$

$$n = m = 1.0$$

$$\Delta\theta_{\text{TO},i} = \Delta\theta_{\text{TO},R}\left(\frac{K_i^2 R + 1}{\text{R} + 1}\right)^n = 31.4°\text{C}$$

$$\Delta\theta_{H,R} = 67.1°\text{C} - 39.0°\text{C} = 28.1°\text{C from the test report}$$

$$\Delta\theta_{H,i} = \Delta\theta_{H,R} \times K_i^{2m} = 21.0°\text{C}$$

$$K_U = \frac{450 \text{ MVA}}{405 \text{ MVA}} = 1.111$$

$$\Delta\theta_{\text{TO},U} = \Delta\theta_{\text{TO},R}\left(\frac{K_U^2 R + 1}{R + 1}\right)^n = 46.0°\text{C}$$

$$\Delta\theta_{H,U} = \Delta\theta_{H,R} \times K_U^{2m} = 34.7°\text{C}$$

The thermal time constant is calculated next (for the FOA cooling class):

$$C = 0.06 \times W_{\text{CC}} + 0.06 \times W_{\text{TF}} + 1.93 \times \text{GO} = 50{,}492 \text{ Wh/°C}$$

$$P_{T,R} = 1{,}190{,}420 \text{ W from the test report}$$

$$\tau_{\text{TO},R} = C \times \Delta\theta_{\text{TO},R}/P_{T,R} = 1.65 \text{ h}$$

Since $n = 1.0$,

$\tau_{\text{TO}} = \tau_{\text{TO},R} = 1.65$ h

Calculating $\Delta\theta_{\text{TO}}$ at $t = 2$ h:

$\Delta\theta_{\text{TO}} = (\Delta\theta_{\text{TO},U} - \Delta\theta_{\text{TO},i})(1 - e^{-t/\tau_{\text{TO}}}) + \Delta\theta_{\text{TO},i} = 41.7°\text{C}$

Calculating $\Delta\theta_{\text{H}}$ by assuming the winding time constant is zero:

$\Delta\theta_H = \Delta\theta_{H,U} = 34.7°\text{C}$

The final result:

$\theta_H = \theta_A + \Delta\theta_{\text{TO}} + \Delta\theta_H = 106.4°\text{C}$ at 2 h

Example 3.7

Using the results from Example 3.6, what are the per unit life and the F_{AA} if 450 MVA are applied *continuously*, assuming a constant 30°C ambient temperature?

The final steady-state hot-spot temperature $\theta_{H,U}$ is based on the ultimate temperature rise of the oil and the ultimate hot-spot gradient:

$$\begin{aligned} \theta_{H,U} &= \theta_A + \Delta\theta_{\text{TO},U} + \Delta\theta_{H,U} \\ &= 30°\text{C} + 46.0°\text{C} + 34.7°\text{C} = 110.7°\text{C} \\ \text{Per unit life} &= 9.8 \times 10^{-18}\, e^{15000/(\theta_{H,U}+273)} = 0.93 \\ F_{\text{AA}} &= e^{39.16-15000/(\theta_{H,U}+273)} = 1.07 \end{aligned}$$

At the 450 MVA loading, the transformer loses life at a rate 7% greater than at the design temperature of 110°C. To calculate the thermal capability precisely, a trial-and-error approach is used where the MVA loading is varied and the temperatures are calculated until the desired per unit life is achieved.

Example 3.8

Three of the transformers shown on the test report shown in Figure 3.16 are used as a generator step-up transformer bank with the HV windings set to the nominal taps. The generator is rated 1280 MVA at 24,000V with a positive-sequence reactance X_{G1} of 0.35 per unit and a negative-sequence reactance X_{G2} of 0.27 per unit. The Y-connected generator is grounded through a resistance. A bolted phase-to-ground fault occurs at one of the HV terminals of the transformer. Using the transformer impedance values shown in the "Average'' column of the transformer test report, calculate the currents at the HV fault and at the generator terminals.

The three-phase diagram of the actual circuit is shown in Figure 3.18. The first step is to convert the generator and transformer impedances to a common MVA and voltage base. In this example, we will use the generator ratings for the common MVA base. The MVA base on a per-phase basis is 1280 MVA/3 = 426. 7 MVA per phase. Let us assume that the series winding impedance of the transformer bank is purely reactive (all leakage reactance). Convert the 14.17% transformer reactance to the new base:

$$X_T = 0.1417 \times \frac{426.7 \text{ MVA}}{405 \text{ MVA}} \times \left(\frac{23{,}600 \text{ V}}{24{,}000 \text{ V}}\right)^2 = 0.1444 \text{ per unit}$$

Load currents are generally neglected in fault calculations. The positive phase sequence, negative phase sequence and zero phase sequence networks are connected in series for a phase-to-ground fault on A phase, as shown in

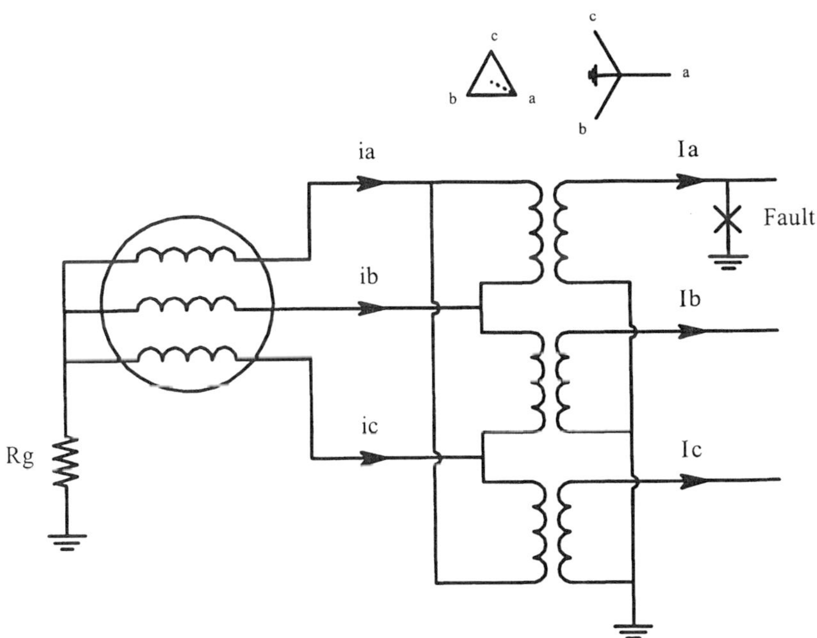

FIGURE 3.18 A phase-to-ground fault at the output of the generator step-up transformer used in Example 3.8.

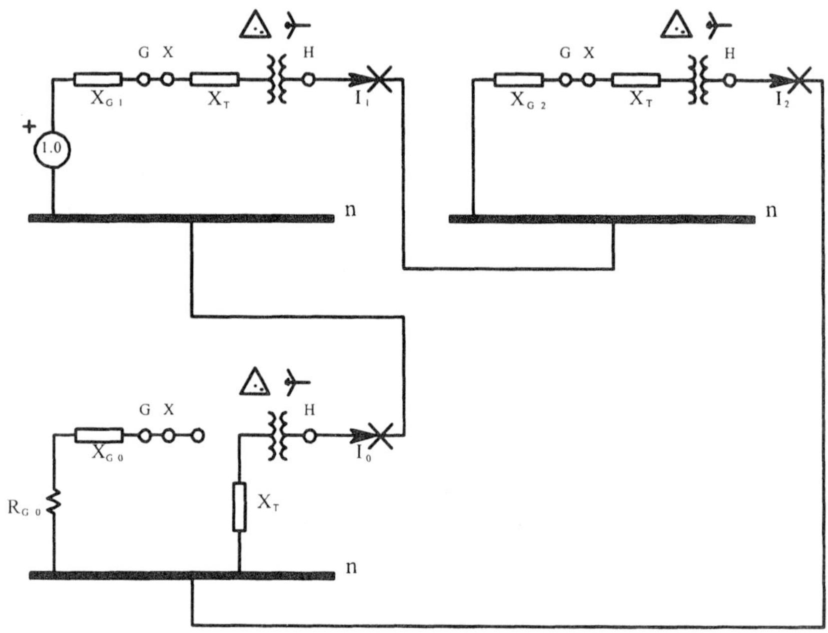

FIGURE 3.19 Sequence network connections for a phase-to-ground fault at the output of a generator step-up transformer used in Example 3.8.

Figure 3.19. Note that although the phase angle of the generator is $-30°$ in the real system, the phase angle of the generator as observed by the HV fault is phase shifted by $+30°$ because of the standard Δ-Y connection with the HV side leading the LV side. Thus, the phase angle of the generator is 0° in the positive-sequence Thévenin equivalent circuit. Note also that the generator grounding resistance and the generator zero-sequence impedance are irrelevant to the calculation since there is no ground current through the transformer secondary circuit because of the Δ connection.

Calculating the HV sequence currents,

$$I_0 = I_1 = I_2 = \frac{1.0}{jX_{G1} + jX_{G2} + 3jX_T} = \frac{1}{j0.35 + j0.274 + j0.4332}$$
$$= 0.946\angle -90° \text{ per unit}$$

Converting the HV sequence currents into phase currents,

$$I_A = I_0 + I_1 + I_2 = 3 \times 0.946\angle -90° \text{ per unit}$$

$$= 2.838\angle{-90°} \text{ per unit}$$

$$I_B = I_0 + I_1\angle 240° + I_2\angle 120° = 0$$

$$I_C = I_0 + I_1\angle 120° + I_2\angle 240° = 0$$

After the voltage base of the LV system is chosen, the voltage base of the HV system must be determined by the transformer voltage ratio and cannot be chosen arbitrarily.

Determining the voltage base of the HV system:

$$V_{\text{base}} = 24{,}000 \text{ V} \times (512{,}500 \text{ V}/23{,}600 \text{ V}) = 521{,}186 \text{ V}$$

Converting $I_{a,b,c}$ into amperes,

$$I_{\text{base}} = 1.732 \times 426{,}000{,}000 \text{ VA}/521{,}186 \text{ V} = 1416 \text{ A per phase}$$

$$I_a = 2.838\angle{-60°} \text{ per unit} \times 1416 \text{ A} = 4019\angle{-90°} \text{ A}$$

$$I_b = I_c = 0$$

Transforming the HV current into LV currents at the generator:

$$i_a = I_a \times (296{,}000 \text{ V}/23{,}600 \text{ V}) = 50{,}408\angle{-90°} \text{ A}$$

$$i_b = -i_a = 50{,}408\angle 90° \text{ A}$$

$$i_c = 0$$

Note that the transformer turns ratio is 296,000/23,600 and *not* 512,500/23,600 in the last step because the ratio of the *winding* voltages must be used instead of the ratio of *system* voltages when transforming voltages and currents from the primary to the secondary.

Example 3.9

A T-connected transformer is used to supply a balanced two-phase secondary load from a three-phase supply. The two-phase load current is 1 per unit in each secondary winding. Calculate the positive, negative, and zero phase sequence currents in the three-phase circuit in per unit.

The T connection uses a 1:1 transformer between *A* phase and neutral to supply one-half of the two-phase load and a 1.732:1 transformer between *C* phase and *B* phase to supply the other half of the two-phase load. The 1:1 transformer has a load current of $1\angle 90°$ on the secondary side and the 1.732:1 transformer has a load current of $1\angle 0°$ on the secondary side. On the three-phase primary side the *A* phase current is $1\angle 90°$, the *B* phase current is $0.577\angle 0°$, and the *C* phase current is $0.577\angle 180°$.

Converting the phase currents into sequence currents:

Zero sequence: $i_0 = \dfrac{i_a + i_b + i_c}{3} = 0.333\angle 90°$

Positive sequence: $i_1 = \dfrac{i_a + i_b\angle 120° + i_c\angle 240°}{3} = 0.666\angle 90°$

Negative sequence: $i_2 = \dfrac{i_a + i_b\angle 240° + i_c\angle 120°}{3} = 0$

Example 3.10

A test circuit for a 2000 KVA three-phase 12470Grd.Y/7200 – 208Grd.Y/120V transformer is shown in Figure 3.20. The ammeter reads 0.2 A and the voltmeter reads 120 V. Calculate the open-circuit zero phase sequence impedance of this transformer, in per unit of the transformer rating.

Full load current per phase at 7200 V:

$$I_{FL} = 2{,}000{,}000\ \text{VA}/(3\ \text{phases} \times 7200\ \text{V}) = 92.6\ \text{A/phase}$$

Zero phase sequence current at 120 V:

$$I_0 = 0.2\ \text{A}/3 = 0.0667\ \text{A/phase}$$

$$Z_0 = E/I = (120\ \text{V}/7200\ \text{V}) \times (92.6\ \text{A}/0.0667\ \text{A}) = 23.1\ \text{per unit}$$

This is essentially the same as Z_{shunt} in the equivalent zero-sequence circuit.

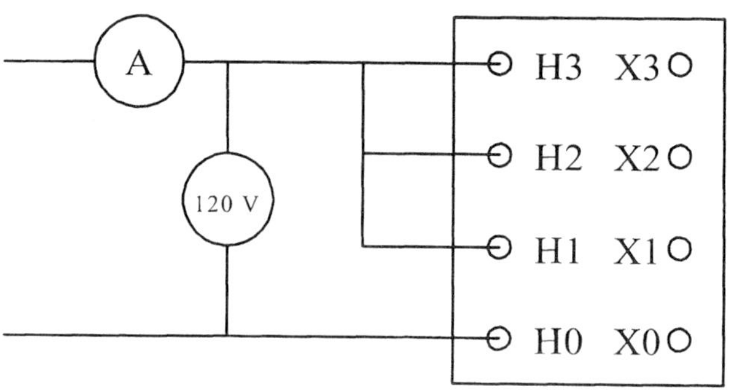

FIGURE 3.20 Test connections used in Example 3.10.

Example 3.11

Three single-phase transformers are connected as a Y-Δ transformer bank. The transformers have the following nameplate ratings:

Transformer *A*:	100 KVA	7200–480	5% Z
Transformer *B*:	100 KVA	7200–480	6% Z
Transformer *C*:	150 KVA	7200–480	5% Z

A 300 KVA balanced three-phase load is applied to the Δ-connected output. Calculate the per-unit load in each transformer on its own base.

Using the method described in Section 3.10:

$$w = \frac{Z_b}{Z_a} = \frac{0.06}{0.05} = 1.2$$

$$x = \frac{Z_c}{Z_a} = \frac{0.05/150\ \text{KVA}}{0.05/100\ \text{KVA}} = 0.667$$

The per unit currents solved for the Δ-connected secondary must be multiplied by 1.732 to be expressed in per-unit loads on a 100 KVA base. Additionally, the current in transformer *C* must be divided by 1.5 to express the per unit load on a 150 KVA base.

$$\begin{aligned} i_a &= 1.732 \times \frac{w\angle 0^\circ - x\angle 120^\circ}{1 + w + x} \\ &= 0.989\angle -20.6^\circ \text{ per unit of 100 KVA} \\ i_b &= 1.732 \times \frac{x\angle 240^\circ - 1\angle 0^\circ}{1 + w + x} \\ &= 0.878\angle 203.4^\circ \text{ per unit of 100 KVA} \\ i_c &= 1.732 \times \frac{1\angle 120^\circ - w\angle 240^\circ}{1 + w + x/1.5} \\ &= 0.770\angle 87.0^\circ \text{ per unit of 150 KVA} \end{aligned}$$

Example 3.12

A 150 MVA Class FOA transformer has the following test data:

Copper losses at full load	940,000 W
Core losses at 100% voltage	150,000 W
Oil temperature rise	39.0°C
HV hot-spot temperature rise	68.1°C
LV hot-spot temperature rise	59.1°C

Using the IEEE method, calculate the aging acceleration factor F_{AA} for this transformer for a constant load of 165 MVA with an ambient temperature of 25°C. From transformer test data:

$$\Delta\theta_{TO,R} = 39.0°C$$

$$\Delta\theta_{H,R} = 68.1°C - 39.0°C = 29.1°C$$

$$K_U = 165 \text{ MVA}/150 \text{ MVA} = 1.1$$

$$R = 940{,}000 \text{ W}/150{,}000 \text{ W} = 6.23$$

$$n = 0.8 \text{ for FOA cooling class}$$

$$m = 0.9 \text{ for FOA cooling class}$$

$$\Delta\theta_{TO,U} = \Delta\theta_{TO,R} \times \left(\frac{K_U^2 R + 1}{R + 1}\right)^n$$

$$= 39.0°C \left(\frac{1.1^2 \times 6.23 + 1}{6.23 + 1}\right)^{0.8}$$

$$= 44.6\ °C$$

$$\Delta\theta_{H,U} = \Delta\theta_{H,R} K_U^{2m} = 29.1°C \times 1.1^{1.8} = 34.5°C$$

$$\theta_A = 25°C$$

$$\theta_H = \theta_A + \Delta\theta_{TO,U} + \Delta\theta_{H,U} = 25.0°C + 44.6°C + 34.5°C = 104.1°C$$

$$F_{AA} = e^{39.16 - 15000/(\theta_H + 273)} = 0.54$$

Example 3.13

Three single-phase transformers are connected in a Δ-Grd.Y generator step-up transformer bank with the generator connected to the Δ side of the bank. Two of the transformers each have a 15% impedance, and the third transformer has a 20% impedance on a common base. A balanced load current of 1 per unit with a 90% lagging power factor is applied at the output of the transformer. What are the positive-, negative-, and zero-sequence voltages at the generator?

The primary of the transformer is modeled as three ideal transformers with series impedances of j0.15, j0.15 for *B* and *C* phases, and j0.20 for *A* phase. The voltages across the windings are E_a, E_b, and E_c. The voltages at the HV terminals are by definition $1\angle 0°$, $1\angle 240°$, and $1\angle 120°$ with the 1 per unit currents 25.8° lagging with respect to the voltages. Refer to Figure 3.21.

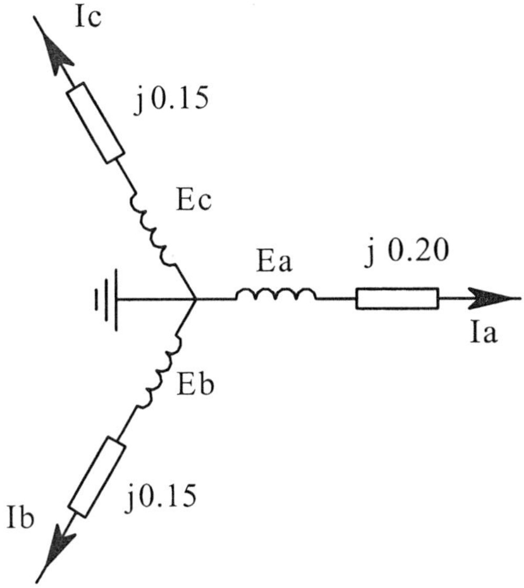

FIGURE 3.21 Diagram used in Example 3.13.

$$E_a = 1\angle 0° + 0.2\angle 64.2° = 1.102\angle 9.4°$$
$$E_b = 1\angle 240° + 0.15\angle 304.2° = 1.074\angle 247.2°$$
$$E_c = 1\angle 120° + 0.15\angle 184.2° = 1.074\angle 127.2°$$

Converting to sequence voltages:

Zero sequence: $E_0 = \dfrac{E_a + E_b + E_c}{3} = 0.0168\angle 64.5°$

Positive sequence: $E_1 = \dfrac{E_a + E_b\angle 120° + E_c\angle 240°}{3} = 1.083\angle 7.9°$

Negative sequence: $E_2 = \dfrac{E_a + E_b\angle 240° + E_c\angle 120°}{3} = 0.0168\angle 64.5°$

The zero-sequence voltage is blocked by the Δ connection on the generator side of the transformer. The positive-sequence and negative-sequence voltages are transformed by the Δ-Grd.Y bank, so *at the generator* these voltages are:

Zero sequence	0
Positive sequence	1.083 per unit
Negative sequence	0.0168 per unit

REFERENCES

1. National Electrical Manufacturers Association. Transformers, Regulators and Reactors. NEMA Standards Publication No. TR 1-1980, Part 7, pp. 3–5.
2. Institute of Electrical and Electronics Engineers, Inc. An American National Standard, IEEE Guide for Loading Mineral-Oil-Immersed Power Transformers Up to and Including 100 MVA with 55°C or 65°C Average Winding Rise. IEEE Std. C57.92-1981, pp. 68–69.
3. Institute of Electrical and Electronics Engineers, Inc. An American National Standard, IEEE Guide for Loading Mineral-Oil-Immersed Power Transformers Up to and Including 100 MVA with 55°C or 65°C Average Winding Rise. IEEE Std. C57.92-1981, pp. 10–11.

4

Autotransformers and Three-Winding Transformers

In this chapter, the principles of the two-winding transformer are extended and applied to two special and very important types of transformers: the autotransformer and the three-winding transformer.

4.1 AUTOTRANSFORMER CONNECTIONS

The autotransformer is both the most simple and the most fascinating of the connections involving two windings. It is used quite extensively in bulk power transmission systems because of its ability to multiply the effective KVA capacity of a transformer. Autotransformers are also used on radial distribution feeder circuits as voltage regulators. The connection is shown in Figure 4.1

The autotransformer shown in Figure 4.1 is connected as a *boosting autotransformer* because the series winding boosts the output voltage. Care must be exercised when discussing "primary" and "secondary" voltages in relationship to windings in an autotransformer. In two-winding transformers, the primary voltage is associated with the primary winding, the secondary voltage is associated with the secondary winding, and the primary voltage is normally considered to be greater than the secondary voltage. In the case of

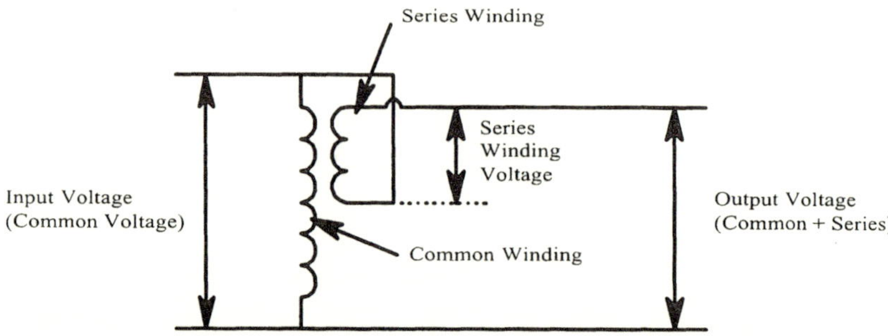

FIGURE 4.1 The boosting autotransformer connection. The output terminals operate at a higher voltage than the input terminals.

a boosting autotransformer, however, the primary (or high) voltage is associated with the series winding, and the secondary (or low) voltage is associated with the common winding; but the voltage across the common winding is higher than across the series winding.

The other possible connection for an autotransformer is shown in Figure 4.2. The autotransformer shown in Figure 4.2 is connected as a *bucking autotransformer* because the series winding bucks, or opposes, the output voltage.

The key feature of an autotransformer is that the KVA throughput of the transformer, i.e., its capacity, is different than the KVA transformed by the common and series windings. The common and series windings are wound on the same core leg, so the transformer laws presented in Chapter 1 still apply:

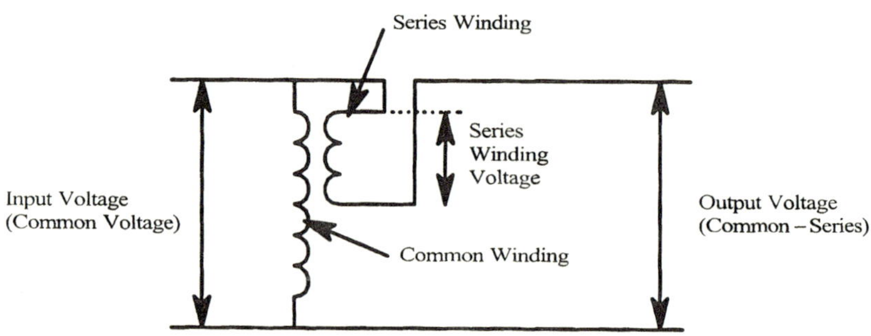

FIGURE 4.2 The bucking autotransformer connection. The output terminals operate at a lower voltage than the input terminals.

1. The volts per turn in the common winding equal the volts per turn in the series winding. The common winding voltage divided by the series winding voltage is equal to the number of turns in the common winding divided by the number of turns in the series winding.
2. The sum of the ampere-turns of the common winding plus the ampere-turns of the series winding equal the magnetizing ampere-turns. The magnetizing ampere-turn are practically zero, so the magnitude of the ampere-turns in the common winding is approximately equal to magnitude of the ampere-turns in the series winding. The series winding current divided by the common winding current is equal to the number of turns in the common winding divided by the number of turns in the series winding.
3. The KVA transformed in the series winding equals the KVA transformed in the common winding.

The capacity multiplication effect stems from the fact that the metallic connection between the input and output circuits allows part of the KVA to flow though the connection and bypass the transformation. This is illustrated in the following example.

Example 4.1

A boosting autotransformer has a common winding voltage of 7200 V and a series winding voltage of 1400 V. The current low-voltage input current is 100 A. Determine the KVA throughput and the KVA transformed. Refer to Figure 4.3.

$$\begin{aligned}
\text{KVA}_{\text{throughput}} &= 7.2\ \text{kV} \times 100\ \text{A} = 720\ \text{KVA} \\
\text{KVA}_{\text{common}} &= \text{KVA}_{\text{series}} = \text{KVA}_{\text{transformed}} \\
7200\ \text{V} \times (100\ \text{A} - I_0) &= 1400\ \text{V} \times I_0 \\
I_0 &= \frac{720\ \text{KVA}}{8600\ \text{V}} = 83.72\ \text{A} \\
\text{KVA}_{\text{common}} &= 7.2\ \text{kV} \times (100\ \text{A} - 83.72\ \text{A}) \\
&= 117.2\ \text{KVA} \\
\text{KVA}_{\text{series}} &= 1.4\ \text{kV} \times 83.72\ \text{A} = 117.2\ \text{KVA}
\end{aligned}$$

In Example 4.1, the ratio of KVA throughput to KVA transformed is 720/117.2 = 6.1, meaning that this autotransformer has 6.1 times the capacity of a two-winding transformer of a similar size and weight. This is a considerable multiplication of KVA capacity.

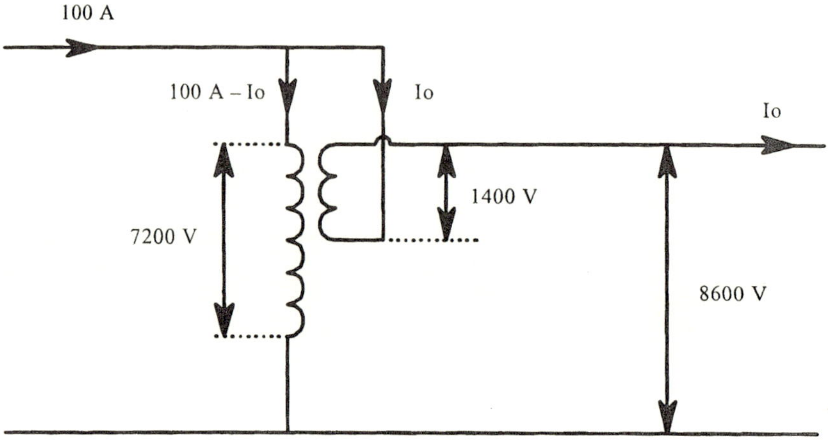

FIGURE 4.3 A boosting autotransformer used in Example 4.1.

Example 4.2

Repeat Example 4.1 for the same windings using a bucking connection instead of a boosting connection with the same 100 A input current. Refer to Figure 4.4

$$\mathrm{KVA}_{\text{throughput}} = 7.2\ \mathrm{kV} \times 100\ \mathrm{A} = 7200\ \mathrm{KVA}$$

$$\mathrm{KVA}_{\text{common}} = \mathrm{KVA}_{\text{series}} = \mathrm{KVA}_{\text{transformed}}$$

$$7200\ \mathrm{V} \times I_c = 1400\ \mathrm{V} \times (100 + I_c) \quad \rightarrow \quad I_c = 24.14\ \mathrm{A}$$

$$\mathrm{KVA}_{\text{common}} = 7.2\ \mathrm{kV} \times 24.14\ \mathrm{A} = 173.8\ \mathrm{KVA}$$

$$\mathrm{KVA}_{\text{series}} = 1.4\ \mathrm{kV} \times (100\ \mathrm{A} + I_c) = 173.8\ \mathrm{KVA}$$

In Example 4.2, the ratio of KVA throughput to KVA transformed is 720/173.8 = 4.1 so the capacity multiplication factor for the bucking connection is not quite as great as that of the boosting connection.

The capacity multiplication factor F_c is a function of the ratio of the output voltage to the input voltage r. For a boosting transformer:

$$F_c = \frac{r}{r - 1} \tag{4.1.1}$$

and for a bucking autotransformer:

$$F_c = \frac{1}{r - 1} \tag{4.1.2}$$

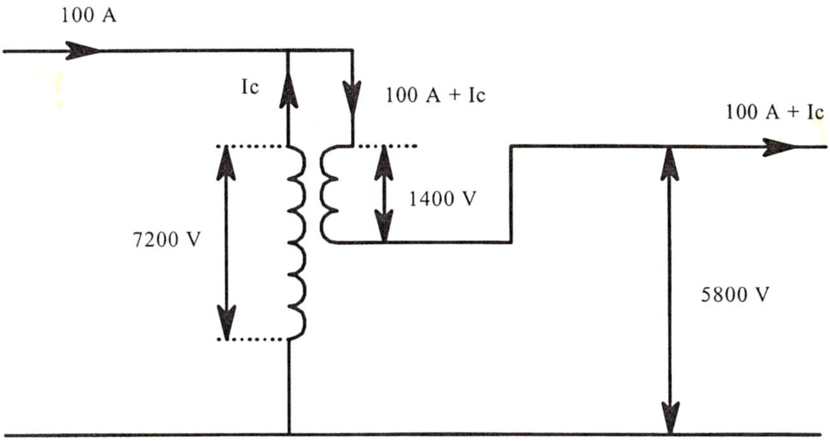

FIGURE 4.4 A boosting autotransformer used in Example 4.2.

For boosting autotransformers and bucking autotransformers, F_c is very large when $r \approx 1$. F_c for a boosting autotransformer is always greater than 1, and it is always greater than F_c for a bucking autotransformer for a given value of r. The chart shown in Figure 4.5 plots F_c versus r for the boosting and bucking connections.

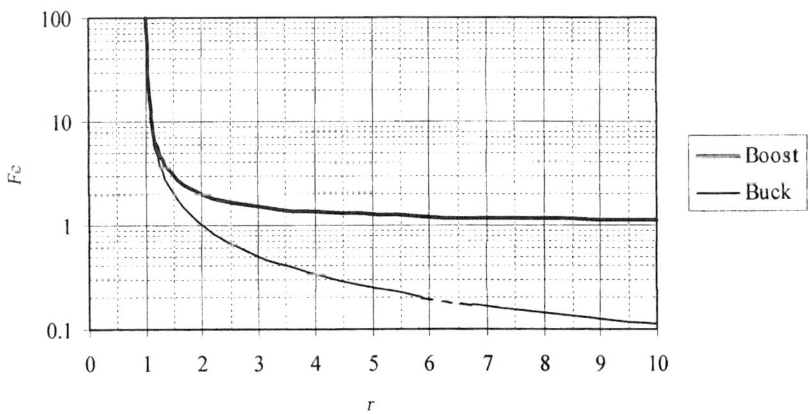

FIGURE 4.5 The capacity factor of an autotransformer F_c as a function of the voltage transformation ratio r. The heavy curve is for a boosting autotransformer and the lighter curve is for a bucking autotransformer.

The bucking connection is never used unless $r \approx 1$ because F_c falls off very rapidly as r increases. A boosting autotransformer begins to lose its advantage over a two-winding transformer as the voltage ratio increases. For $r > 10$, there is practically no advantage at all.

4.2 IMPEDANCE OF AN AUTOTRANSFORMER

The impedance of a transformer can be determined by measuring the impedance across the primary terminals with the secondary terminals short-circuited. Consider a two-winding transformer having a turns ratio n connected as a boosting autotransformer. The transformer impedance, which is mainly leakage reactance, is split between the common winding and the series winding as Z_c and Z_s, respectively. The magnetizing impedance is neglected. To determine the impedance of the autotransformer, the secondary (low-voltage) terminals are short-circuited and a voltage source E_p is applied to the primary (high-voltage) terminals, as shown in Figure 4.6.

The current through the common winding I_c is equal to the current through the series winding divided by the turns ratio:

$$I_c = \frac{I_s}{n} \tag{4.2.1}$$

The voltage across the common winding E_c is equal to the voltage drop across Z_c:

$$E_c = I_c \times Z_c = I_s \times \frac{Z_c}{n} \tag{4.4.2}$$

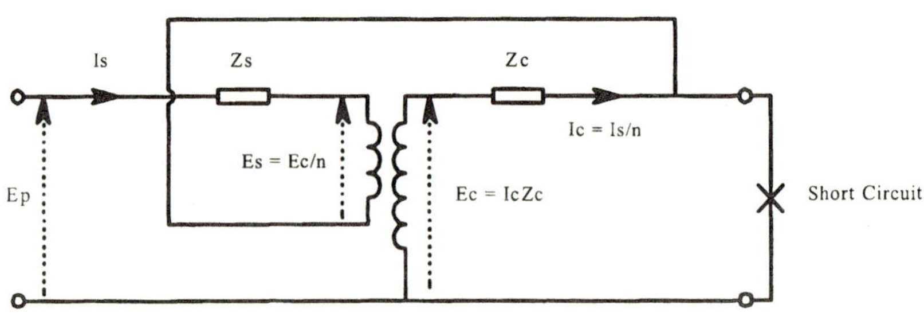

FIGURE 4.6 A boosting autotransformer with a short circuit applied at the low-voltage output.

The voltage across the series winding E_s is equal to E_c divided by the turns ratio:

$$E_s = \frac{E_c}{n} = I_s \times \frac{Z_c}{n^2} \tag{4.2.3}$$

The current through the series winding I_s is equal to the difference between the input voltage E_p and the voltage across the series winding E_s divided by Z_s:

$$I_s = \frac{E_p - E_s}{Z_s} \tag{4.2.4}$$

Multiplying both sides of Eq. (4.2.4) by Z_s and substituting Eq. (4.2.3) for E_s:

$$I_s \times Z_s = E_p - I_s \times \frac{Z_c}{n^2} \tag{4.2.5}$$

Solving for E_p by rearranging Eq. (4.2.5):

$$E_p = I_s \times \left(Z_s + \frac{Z_c}{n^2}\right) \tag{4.2.6}$$

The impedance as seen at the primary or high-voltage side of the transformer Z_p is equal to E_p/I_s. This is obtained by dividing both sides of Eq. (4.2.6) by I_s:

$$Z_p = Z_s + \frac{Z_c}{n^2} \tag{4.2.7}$$

Notice that the value of Z_p in Eq. (4.2.7) is equal to the series impedance as seen at the *secondary* or *low-voltage* side when the transformer is connected as a conventional two-winding transformer with the high-voltage windings short circuited. For a large ratio n this impedance is much smaller than the series impedance seen at the primary side.

In general, the series impedance of a small two-winding transformer is greater than the series impedance of a large two-winding transformer if the two impedances expressed as percent of the winding KVA bases are equal. However, when a small transformer is connected as an autotransformer the series impedance seen at the transformer terminals can be substantially smaller than the impedance of a much larger equivalent two-winding transformer. This can be quantified as follows. Let:

$\% Z_{AT}$ = impedance of a unit connected as autotransformer expressed in percent of the winding KVA base
KVA_{AT} = winding KVA base a unit connected as an autotransformer
$\% Z_{TW}$ = impedance of a two-winding transformer expressed in percent of the winding KVA base
KVA_{TW} = winding KVA base of a two-winding transformer
Z_{AT} = impedance, in ohms, of an autotransformer across the primary terminals
Z_{TW} = impedance, in ohms, of a two-winding transformer across the primary terminals
r = ratio of the primary terminal voltage to the secondary terminal voltage
n = turns ratio of the common/series windings of the autotransformer
E_p = primary voltage
E_s = autotransformer series winding voltage
η = ratio of Z_{AT}/Z_{TW}

The impedance of a transformer in ohms is equal to the per unit impedance times the quantity E^2_{base}/KVA_{base}. Recall from Eq. (4.2.7) that the impedance of an autotransformer seen at the primary terminals is equal to the series impedance as seen at the *secondary* or *low-voltage* side when the transformer is connected as a conventional two-winding transformer. Therefore, $E_{base} = E_s$, and

$$Z_{AT} = 0.01 \times \%Z_{AT} \times E_s^2/KVA_{AT} \qquad \text{ohms} \tag{4.2.8}$$

For a two-winding transformer, $E_{base} = E_p$, and

$$Z_{TW} = 0.01 \times \%\ Z_{TW} \times \frac{E_p^2}{KVA_{TW}} \qquad \text{ohms} \tag{4.2.9}$$

The ratio Z_{AT}/Z_{TW} is found by dividing Eq. (4.2.8) by Eq. (4.2.9).

$$\frac{Z_{AT}}{Z_{TW}} = \frac{E_s^2 \times KVA_{TW} \times \%Z_{AT}}{E_p^2 \times KVA_{AT} \times \%Z_{TW}} \tag{4.2.10}$$

If the two-winding transformer is replaced by an autotransformer having the same KVA capacity, the KVA rating of the autotransformer windings is smaller than the KVA rating of the two-winding transformer windings. Recalling the capacity multiplication factor of a boosting autotransformer in Eq. (4.1.1),

$$\mathrm{KVA_{TW}/KVA_{AT}} = F_c = \frac{r}{r - 1} \tag{4.2.11}$$

For the boosting autotransformer connection, the secondary voltage is equal to the voltage across the common winding. The voltage across the series winding E_s divided by the secondary voltage is therefore equal to $1/n$:

$$\frac{E_s}{\text{secondary voltage}} = \frac{1}{n} \tag{4.2.12}$$

The ratio r was defined as the primary terminal voltage divided by the secondary terminal voltage:

$$\frac{E_p}{\text{secondary voltage}} = r \tag{4.2.13}$$

For a boosting autotransformer,

$$\frac{1}{n} = r - 1, \tag{4.2.14}$$

$$\frac{E_s}{E_p} = \frac{r - 1}{r} \tag{4.2.15}$$

$$(E_s/E_p)^2 = \frac{(r - 1)^2}{r^2} \tag{4.2.16}$$

Combining Eqs. (4.2.11) and (4.2.16) with Eq. (4.2.10), the following result is obtained.

$$\eta = \frac{(r - 1) \times \%Z_{AT}}{r\ \%Z_{TW}} \tag{4.2.17}$$

Since $\%Z_{AT}$ and $\%Z_{TW}$ are expressed on the KVA bases of the transformer *windings*, the values of these values should be comparable even though the physical sizes and KVA ratings of the transformers may be very different.

For the special case where $\%Z_{AT}$ and $\%Z_{TW}$ are equal,

$$\eta = \frac{r - 1}{r} < 1 \tag{4.2.18}$$

The meaning of Eq. (4.2.18) is that the engineer has the opportunity to design an autotransformer with much lower conductor losses and regulation than would be practical with a two-winding transformer with the same KVA capacity. However, in order to limit short-circuit currents, the reactance component of the impedance must still be maintained above some minimum design limit. In Section 4.4, we will see that an autotransformer behaves very differently from a two-winding transformer when a short-circuit is applied to the output.

Example 4.3

A certain three-phase autotransformer has a 50,000 KVA capacity and a primary to secondary voltage rating of 230,000/132,791Grd.Y-69,000/39,837Grd.Y. In order to limit fault current to less than 7 per unit, the system design engineer specified a 15% impedance on a 50,000 KVA base. Determine the required $\%Z$ of the windings, based on the KVA rating of the *windings* and not the KVA capacity of the autotransformer, in order to provide the required 15% impedance.

In this instance, impedance of the autotransformer is equal to the impedance of the equivalent two-winding transformer, both being 15% on a 50,000 KVA base and having the same impedance in ohms.

$$Z_{AT} = Z_{TW}$$
$$\eta = 1$$

From Eq. (4.2.17),

$$\eta = \frac{(r - 1) \times \%Z_{AT}}{r \times \%Z_{TW}}$$

$$r = \frac{230{,}000}{69{,}000} = 3.333$$

$$\eta = \left(1 - \frac{1}{r}\right) \times \frac{\%Z_{AT}}{\%Z_{TW}} = 1$$

Solving for $\%\ Z_{AT}$,

$$\%\ Z_{AT} = \frac{\%Z_{TW}}{1 - 1/r} = \frac{15\%}{1 - 0.3} = 21.4\%$$

Example 4.4

Suppose the series impedance of the autotransformer in Example 4.1 is 10% of its winding KVA base. What is the impedance of the autotransformer expressed as a percent of the KVA *capacity* of the autotransformer?

In Example 4.1, the winding KVA base is equal to 117.2 KVA that is *transformed*. The autotransformer's *capacity* is equal to its 720 KVA throughput Equation (4.2.18) gives the impedance (in ohms) of the autotransformer divided by the impedance (in ohms) of an equivalent two-winding transformer if both transformers have the same percent impedance values of their winding KVA bases. This ratio is calculated below.

$$\eta = \frac{r - 1}{r}$$

$$r = \frac{8600 \text{ V}}{7200 \text{ V}} = 1.194$$

$$\eta = \frac{1.194 - 1}{1.194} = 0.162$$

The ratio η given by Eq. (4.2.18) is also equivalent to the ratio of the *percent* impedance of the autotransformer divided by the *percent* impedance of an equivalent two-winding transformer, when both percent impedances are expressed on the same KVA base. In other words, the 117.2 KVA two-winding transformer with a 10% impedance in this example can be connected as a boosting autotransformer and the autotransformer's impedance will be 1.62% of the 117.2 KVA base.

When the autotransformer's impedance is expressed on a 720 KVA base, the resulting impedance is still lower.

$$\%Z = 1.62\% \times = \frac{117.2 \text{ KVA}}{720 \text{ KVA}} = 0.264\%$$

Since transformer load losses can be seen as a resistance component of the series impedance, the load losses of the autotransformer are on the order of 0.00264 times the load losses of a larger two-winding transformer with a similar coil design. This example clearly illustrates the tremendous savings in economy that can be realized by using the autotransformer connection.

Example 4.5

A 7200–120 V two-winding transformer has a 7% impedance based on the winding ratings. The transformer is connected as a single-phase boosting autotransformer in a single-phase radial distribution circuit. What is the % impedance based on the capacity of the autotransformer connection?

The input to the autotransformer is 7200 V single phase. The output from the autotransformer is 7200 V + 120 V = 7320 V single phase.

$$r = \frac{7320}{7200} = 1.01667$$

At the output side of the autotransformer, the impedance of the autotransformer is the impedance of the series winding, or 7% on a 120 V base. Converting to a 7320 V base:

$$\%Z = 7\% \times \left(\frac{120}{7320}\right)^2 = 0.00188\%$$

The KVA capacity of the transformer in the boosting autotransformer connection is increased by the factor $r/(r - 1)$. Converting the impedance to the autotransformer KVA capacity base,

$$\%Z = 0.00188\% \times \frac{r}{r-1} = 0.00188\% \times \frac{1.01667}{0.01667} = 0.115\%$$

4.3 LIMITATIONS OF THE AUTOTRANSFORMER CONNECTION

One of the limitations of the autotransformer connection is that not all types of three-phase connections are possible. For example, the Δ-Y and Y-Δ connections are not possible using the autotransformer. The Y-Y connection must

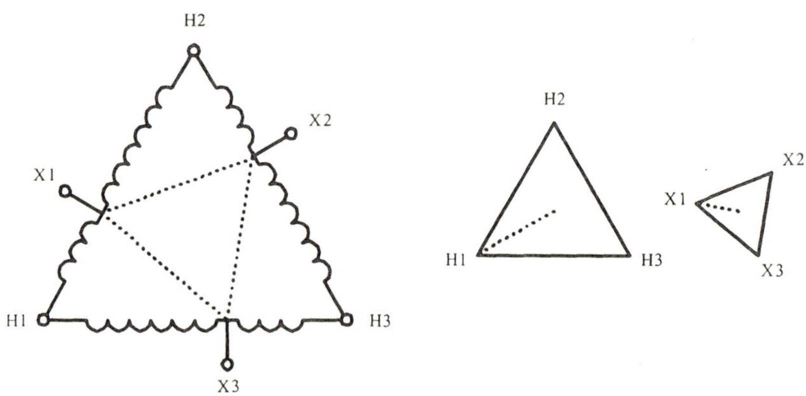

FIGURE 4.7 A hypothetical Δ-Δ autotransformer connection showing the phase angle displacement between the high-voltage and low-voltage terminals. For this reason, this autotransformer connection is seldom, if ever, used.

share a common neutral between the high-voltage and low-voltage windings, so the neutrals of the circuits connected to these windings cannot be isolated. The Grd.Y-Grd.Y connection using autotransformers is indistinguishable from a Grd.Y-Grd.Y connection using two-winding transformers. This is the most common connection using autotransformers.

A Δ-Δ autotransformer connection is theoretically possible; however, this will create a peculiar phase shift. The phase shift is a function of the ratio of the primary to secondary voltages and it can be calculated from the vector diagram, such as the example shown in Figure 4.7. This phase shift cannot be changed or eliminated and for this reason, autotransformers are very seldom connected as Δ-Δ transformers.

4.4 AUTOTRANSFORMER VOLTAGES WITH SHORT CIRCUITS APPLIED

When a short circuit is applied across the secondary terminals of a two-winding transformer, the voltages across both the primary and secondary windings decrease because of exclusion of flux from the core, as was described in Chapter 3. This is easily seen from the equivalent circuit of a two-winding transformer: The voltage across the magnetizing branch is reduced by approximately 50% when the secondary terminals short-circuited due to the voltage drop across the primary series impedance, assuming the per unit primary series impedance Z_H is approximately the same as the per unit secondary series impedance Z_X.

The situation is very different when a short circuit is applied to an autotransformer, as can be seen from the circuit diagram shown in Figure 4.8. All of the applied primary voltage E_p appears across the impedance of the series

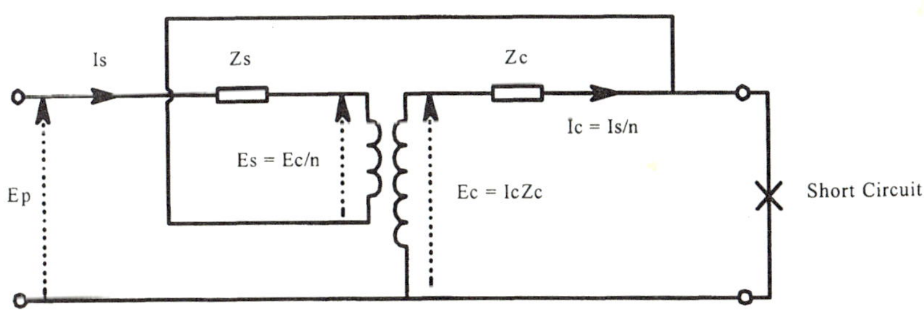

FIGURE 4.8 Measuring the series impedance of an autotransformer involves shorting the output terminals and measuring the current at the input terminals.

winding Z_s so the series current is equal to the applied voltage divided by the impedance:

$$I_s = \frac{E_p}{Z_s} \tag{4.4.1}$$

This current is transformed into current in the common winding:

$$I_c = \frac{I_s}{n} = \frac{E_p}{n \times Z_s} \tag{4.4.2}$$

The voltage across the common winding is equal to the voltage drop through the common winding times the impedance of the common winding Z_c:

$$E_c = I_c \times Z_c = \frac{E_p \times Z_c}{n \times Z_s} \tag{4.4.3}$$

The voltage across the common winding is transformed into a voltage E_s across the series winding.

$$E_s = \frac{E_c}{n} = \frac{E_p \times Z_c}{n^2 \times Z_s} \tag{4.4.4}$$

If the *per-unit* values of Z_c and Z_s are approximately equal, then the ratio $Z_c/(n^2 \times Z_s)$ is approximately equal to 1:

$$E_s \approx E_p \tag{4.4.5}$$

The series winding of an autotransformer has a normal applied voltage that is usually only a small fraction of the applied voltage across the primary terminals. A short circuit across the secondary terminals causes a very significant *increase* in the voltages across the series winding. This almost guarantees that the core will go into saturation, with two immediate results. First, the admittance of the magnetizing branch of the transformer can no longer be ignored, and the admittance of the magnetizing branch is very nonlinear with the core in saturation; therefore, short-circuit current calculations are somewhat uncertain for autotransformers. Second, with the core in saturation, the transformer essentially becomes an air-core transformer, further limiting the ability to calculate impedances. Fortunately, this also limits the magnitude of the induced voltage across the common winding. So although the voltage across the series winding may be several times the normal operating voltage of this winding, the overvoltage across the common winding is not quite as severe.

4.5 IMPULSE VOLTAGES APPLIED TO AUTOTRANSFORMERS

Another feature of the autotransformer needs to be dealt with; namely, when a voltage surge travels along the primary circuit and hits the series winding, nearly all of the surge voltage is applied across the series winding. This is because the relatively large line-to-ground capacitance of the external secondary circuit prevents the voltage to build up at the secondary terminals before the voltage surge has reached its peak at the primary terminals. Because the turns ratio between the common and series winding is usually large, a very high voltage is also induced in the common winding. The ASA Test Code, which was used in the past, recognized this particular disadvantage of autotransformers and required full impulse test voltages to be applied across each winding in turn (see Ref. 1.)

The modern IEEE test code does not make any special provisions for testing autotransformers; however, the design engineer must be cognizant of the fact that significantly higher impulse voltages may appear across the windings of an autotransformer as compared to a two-winding transformer.

One way that this problem can be addressed is to install surge protection in autotransformers. Thyrite blocks are connected between the windings and ground and across the series winding to limit applied and induced impulse voltages. Thyrite blocks have a nonlinear resistance characteristic with the current through the thyrite increasing sharply when the voltage exceeds a certain breakdown voltage. This tends to limit the voltage across the thyrite, thereby protecting the windings. In a boosting autotransformer, thyrite blocks are typically installed across the common winding, between the high-voltage terminals and ground, and across the series winding.

As was seen in Section 4.4, the series and common windings are exposed to excessive voltages during short-circuit conditions as well as during line surges. The problem with using thyrite blocks or other types of surge protection in autotransformers is that the surge protection may break down under short-circuit conditions unless their breakdown voltages are carefully coordinated.

4.6 AUTOTRANSFORMER CORE AND COIL DESIGNS AND TERMINAL CONFIGURATIONS

Engineers sometimes conceptualize each phase of an autotransformer as a single winding with a tap brought out and connected to the secondary circuit.

This conceptualization is electrically correct but not physically accurate, however. The voltages and currents of the series and common windings are usually very different, so the two windings are usually designed very differently. For example, if the series winding has fewer turns than the common winding, as is usually the case, the series winding will carry more current and the conductors must have a larger cross-sectional area than the common winding. In very large autotransformers the series winding may even be constructed as multiple parallel windings.

So a typical core and coil assembly in an autotransformer will be somewhat similar to the core and coil assembly of a two-winding transformer with the common and series windings being concentric and wound around the same core leg. However, unlike a two-winding transformer, the series winding will be located farthest from the core since it operates at the line potential of the high-voltage circuit. Also, the series winding will be insulated for much higher phase-to-ground voltages as well as much higher turn-to-turn voltages than the secondary of a two-winding transformer in order to withstand the high voltages across it due to line surges and secondary short circuits.

For three-phase autotransformers, all of the standard two-winding core and coil designs are used: three-legged core form, five-legged core form, and shell form. The major difference between the autotransformer design and the two-winding design is in the way the windings are interconnected.

Externally, there is little difference between autotransformers and two-winding transformers. The standard terminal arrangements apply to both types of connections. The only difference is that with an autotransformer, the X0 and H0 terminals must be brought out through a common neutral bushing. This neutral bushing is sometimes labeled the H0/X0. Single-phase autotransformers can still be additive or subtractive, and the external locations of the bushing terminals still follow the standard configuration; however, since autotransformers tend to have large KVA ratings, they are usually subtractive polarity.

4.7 ADVANTAGES AND DISADVANTAGES OF THE AUTOTRANSFORMER CONNECTION

Summarizing the advantages of the autotransformer connection:

- There are considerable savings in size and weight.
- There are decreased losses for a given KVA capacity.

- Using an autotransformer connection provides an opportunity for achieving lower series impedances and better regulation.

Summarizing the disadvantages of the autotransformer connection:

- The autotransformer connection is not available with certain three-phase connections.
- Higher (and possibly more damaging) short-circuit currents can result from a lower series impedance.
- Short circuits can impress voltages significantly higher than operating voltages across the windings of an autotransformer.
- For the same voltage surge at the line terminals, the impressed and induced voltages are greater for an autotransformer than for a two-winding transformer.

In many instances, the advantages of the autotransformer connection outweigh its disadvantages. For example, when very large MVA capability is required and where a Grd.Y-Grd.Y connection is suitable, an autotransformer is usually the design of choice. Because an autotransformer cannot provide a Δ-Y connection, autotransformers are not suitable for use as generator step-up transformers.

4.8 THREE-WINDING TRANSFORMERS

The three-winding transformer is a subset of multiwinding transformers. In addition to the usual primary and secondary windings, a third *tertiary* winding is added to each phase. Having three winding can serve several purposes:

- Three windings allow connecting three systems together where each system has a different operating voltage.
- Three windings provide electrical isolation between dual input circuits or dual output circuits having the same operating voltage.
- If the third winding is Δ-connected, this can stabilize voltages, supply third harmonic currents to magnetize the transformer core, filter third harmonics from the system, and provide grounding bank action when the primary and secondary windings are both Y-connected.

Sometimes a tertiary winding may serve more than one function at the same time. For example, a 13.8 kV Δ-connected tertiary winding on a 230 kV–69 kV Grd.Y-Grd.Y transformer helps to stabilize the primary and secondary voltages, provides grounding bank action to partially shield the primary circuit from secondary ground currents, and provides 13.8 kV supply voltage to a station-service auxiliary transformer. (*Note*: When a group of windings are connected in parallel to increase the current capability of a secondary winding, the parallel group is considered one winding and not several separate windings. Using multiple sets of low-voltage windings in parallel is common in large generator step-up transformers; however, these are still considered two-winding transformers.)

Sometimes a tertiary winding is intended only to magnetically interact with the primary and secondary windings so it may not have any external terminal connections. In these cases, the tertiary winding is said to be an *imbedded* tertiary. Imbedded tertiary windings are found only in three-phase transformers and are always Δ-connected. One corner of the Δ-connected imbedded tertiary winding is sometimes grounded internally to limit capacitively coupled voltages.

For single-phase transformers, the standard labels for the tertiary bushings are Y1 and Y2. For three-phase transformers, the standard labels for the tertiary bushings are (Y0), Y1, Y2, Y3.

4.9 MODIFICATION OF TRANSFORMER LAWS WITH THREE WINDINGS

The transformer laws developed in Chapter 1 do not fundamentally change with the addition of a tertiary winding to the core leg. Since the tertiary winding is linked to the same flux as the primary and secondary windings, the volts per turn of the tertiary is identical to the volts per turn of the other two windings.

$$\frac{E_H}{N_H} = \frac{E_X}{N_X} = \frac{E_Y}{N_Y} \tag{4.9.1}$$

where

E_H = voltage across the primary winding
E_X = voltage across the secondary winding
E_Y = voltage across the tertiary winding
N_H = number of turns in the primary winding
N_X = number of turns in the secondary winding
N_Y = number of turns in the tertiary winding

There are now three transformer turns ratios instead of one and the first transformer law can be modified as follows:

$$\frac{E_H}{E_X} = \frac{N_H}{N_X} \tag{4.9.2}$$

$$\frac{E_H}{E_Y} = \frac{N_H}{N_Y} \tag{4.9.3}$$

$$\frac{E_X}{E_Y} = \frac{N_X}{N_Y} \tag{4.9.4}$$

When current flows in the tertiary winding, this adds to (or subtracts from, depending on the direction of the current) the total MMF. The total MMF is equal to the magnetizing MMF, which being small is usually assumed to be zero.

$$(i_H \times N_H) + (i_X \times N_X) + (i_Y \times N_Y) = \mathrm{MMF_{mag}} \approx 0 \tag{4.9.5}$$

where

i_H = current in the primary winding
i_X = current in the secondary winding
i_Y = current in the tertiary winding

Since $E_H/N_H = E_X/N_X = E_Y/N_Y$ = constant, per Eq. (4.9.1), each of the terms above can be multiplied by the same constant:

$$\begin{aligned}(i_H \times N_H) \times \left(\frac{E_H}{N_H}\right) &+ (i_X \times N_X) \\ \times \left(\frac{E_X}{N_X}\right) &+ (i_Y \times N_Y) \times \left(\frac{E_Y}{N_Y}\right) = 0\end{aligned} \tag{4.9.6}$$

This equation reduces to the following form, resulting in the modification to the second transformer law:

$$(i_H \times E_H) + (i_X \times E_X) + (i_Y \times E_Y) = 0 \tag{4.9.7}$$

Equation (4.9.7) can be interpreted as an energy conservation law: The total instantaneous power flowing into the transformer from all three windings is zero. In a two-winding transformer, the magnitude of the KVA in the primary winding equals the magnitude of the KVA in the secondary windings; however, in a three-winding transformer, the KVA may divide among the three windings in an infinite number of ways, depending on the combination of loads applied to the secondary and tertiary circuits.

4.10 EQUIVALENT CIRCUIT OF A THREE-WINDING TRANSFORMER

Various forms of a three-winding transformer equivalent circuit have been proposed, but the simplest and most useful is the so-called *T equivalent circuit*, shown in Figure 4.9. The magnetizing branch is omitted in the *T* equivalent since the magnetizing impedance is normally much greater than the series impedances. If voltages and impedances are expressed in per unit values, then the ideal transformers can sometimes be omitted also; however, in some cases 1:1 ideal transformers are retained so that the connections to the primary, secondary and tertiary circuits can be properly represented by the equivalent circuit.

In a three-winding transformer, eddy-current losses occur in each winding from stray flux produced by the other two windings, even if the third winding is not carrying any load. Therefore, each series resistance element in the *T* equivalent circuit of a three-winding transformer represent eddy-current losses produced by currents in other windings. Hence, a series resistance does not belong to any particular winding but is distributed among all three widings.

To derive the series impedance values in the *T* equivalent circuit, impedance measurements are made of each pair of windings taken two at a time. One winding is short-circuited with voltage applied to the other winding while the third winding is open-circuited. The current is measured through the winding with the applied voltage. The impedance is equal to the applied voltage divided by that current. The test setup to measure the impedance between the *H* and *X* windings of a single-phase three-winding transformer is shown in Figure 4.10.

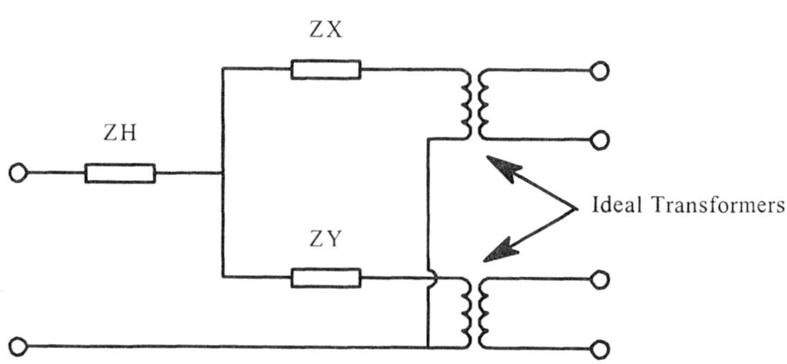

FIGURE 4.9 *T* equivalent circuit of a three-winding transformer.

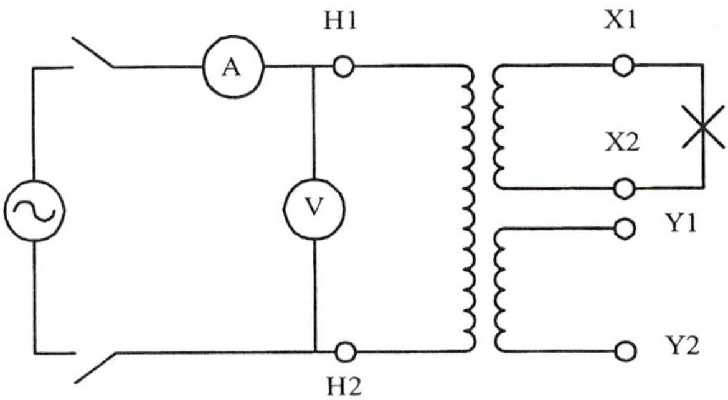

FIGURE 4.10 Measuring Z_{HX} of a three-winding transformer.

The test for a three-phase, three-winding transformer is similar except that three-phase voltages are used. There are three sets of measurements taken. The first set of measurements applies a three-phase voltage to the H1, H2, and H3 terminals with the X1, X2, and X3 terminals shorted together and the Y1, Y2, and Y3 terminals open. The second set of measurements applies a three-phase voltage to the H1, H2, and H3 terminals with the Y1, Y2, and Y3 terminals shorted together and the X1, X2, and X3 terminals open. Finally, a three-phase voltage is applied to the X1, X2 and X3 terminals with the Y1, Y2, and Y3 terminals shorted together with the H1, H2, and H3 terminals open.

The Z_{HX}, Z_{HY}, and Z_{XY} impedance values are determined by dividing the voltages by the currents in each test. The series impedances of the *T* equivalent circuit are calculated using the following equations.

$$Z_H = \frac{Z_{HX} + Z_{HY} - Z_{XY}}{2} \tag{4.10.1}$$

$$Z_X = \frac{Z_{HX} + Z_{XY} - Z_{HY}}{2} \tag{4.10.2}$$

$$Z_Y = \frac{Z_{HY} + Z_{XY} - Z_{HX}}{2} \tag{4.10.3}$$

One of the windings of a three-winding transformer may have a much smaller KVA rating than the other two windings but the series impedance of that winding may have about the magnitude as the series impedances of the other two windings. If the winding having the smallest rating is short-circuited, the current in that winding may be much greater than its ''short-circuit withstand capability'' unless the impedance of that winding is increased. This can be done by physically placing the winding farther away from the other windings to increase the leakage flux. There are practical limitations to doing this, however, so the lowest-rated winding is usually no less than 35% of the rating of the other windings in order to withstand the effects of short circuits.

It is possible for one of the impedances in the *T* equivalent to have a negative impedance value. This occurs if one of the values of Z_{XY}, Z_{HY}, or Z_{HX} is greater than the sum of the other two values. This condition makes one of the values of Z_H, Z_X, or Z_Y negative, but only one of these can be negative in the same transformer. When connecting single-phase three-winding transformers in a bank, it is extremely important that *all* of the impedance values are matched; otherwise, unbalanced operation will result. A branch with a high negative impedance can also affect ground relaying.

Factory tests for a three-phase Y-Y-Δ transformer will usually include zero phase sequence impedance measurement. This measurements is done by jumpering the H1, H2, and H3 bushings together and applying a *single-phase* voltage between these bushings and the H0 bushing while measuring the resulting zero phase sequence current. The value of Z_0 on a per-phase basis is equal to the applied voltage divided by one-third of the total current.

4.11 CORE AND COIL CONSTRUCTION OF THREE-WINDING TRANSFORMERS

The primary, secondary, and tertiary windings of each phase are physically wound on a common core leg. Any of the two-winding core and coil designs are used for three-winding transformers, including single-phase core form; single-phase shell form; three-phase, three-legged and five-legged core form; and three-phase shell form designs. In core-form designs, it is customary to place the winding with the lowest operating voltage nearest to the grounded core, with the intermediate-voltage winding wound around it, and the high-voltage winding wound around the intermediate-voltage winding. Since the tertiary winding is often the winding with the lowest operating voltage, it would follow that the tertiary winding would be closest to the core. Because it is sometimes necessary to design the tertiary winding with a high series impedance to limit the short-circuit current, occasionally the tertiary is wound around the high-voltage and low-voltage windings to increase the leakage flux for this winding and create a high leakage reactance between the tertiary and the other two windings. The placement of the windings will measurably affect the distribution of impedances in the *T* equivalent circuit.

Example 4.6

Three single-phase transformers are connected in a Grd.Y-Grd.Y-Δ transformer bank. The primary and secondary windings both have the same MVA rating and are both connected Grd.Y. The tertiary winding MVA ratings are 35% of the primary ratings. The transformers have the following per unit impedances.

Transformers *A* and *B*:	$Z_{HX} = 0.14$	$Z_{HY} = 0.35$	$Z_{XY} = 0.20$
Transformer *C*:	$Z_{HX} = 0.14$	$Z_{HY} = 0.20$	$Z_{XY} = 0.35$

All impedances are assumed to be purely reactive and are expressed in per unit of the MVA base of the primary and secondary windings. There is no external load on the tertiary circuit and the primary voltages are balanced. Describe what happens when balanced three-phase load currents are applied to the secondary windings of this transformer bank.

Calculating the *HV*, *LV*, and *TV* impedances for transformers *A* and *B*:

$$Z_H = \frac{Z_{HX} + Z_{HY} - Z_{XY}}{2} = \frac{0.14 + 0.35 - 0.20}{2} = 0.145 \text{ per unit}$$

$$Z_X = \frac{Z_{HX} + Z_{XY} - Z_{HY}}{2} = \frac{0.14 + 0.20 - 0.35}{2} = -0.005 \text{ per unit}$$

$$Z_Y = \frac{Z_{HY} + Z_{XY} - Z_{HX}}{2} = \frac{0.35 + 0.20 - 0.14}{2} = 0.410 \text{ per unit}$$

Calculating the *HV*, *LV*, and *TV* impedances for transformer C:

$$Z_H = \frac{Z_{HX} + Z_{HY} - Z_{XY}}{2} = \frac{0.14 + 0.20 - 0.35}{2} = -0.005 \text{ per unit}$$

$$Z_X = \frac{Z_{HX} + Z_{XY} - Z_{HY}}{2} = \frac{0.14 + 0.35 - 0.20}{2} = 0.145 \text{ per unit}$$

$$Z_Y = \frac{Z_{HY} + Z_{XY} - Z_{HX}}{2} = \frac{0.35 + 0.20 - 0.14}{2} = 0.410 \text{ per unit}$$

The *T* equivalent circuit of the transformer bank in Example 4.6 is shown in Figure 4.11. Because the position of the negative-impedance leg for transformer *C* is reversed in its *T* equivalent circuit, a voltage unbalance is produced when currents flow. The voltage unbalance is proportional to the load current. This causes a circulating current I_y to flow through in the tertiary windings.

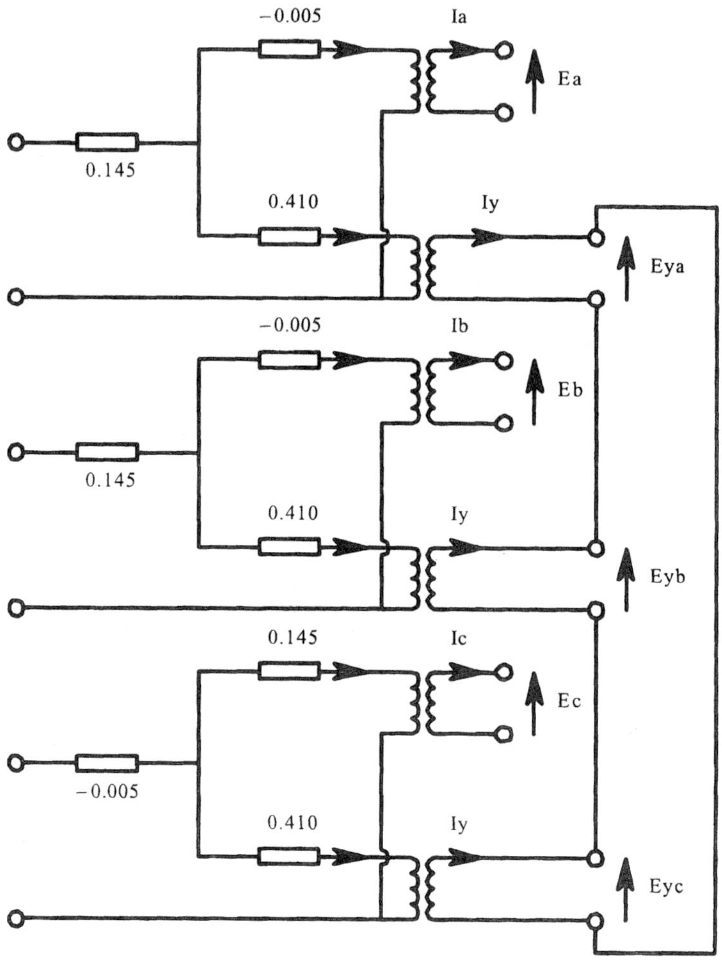

FIGURE 4.11 Bank of three-winding transformers with unbalanced series impedances used in Example 4.6.

For 1.0 per unit balanced load currents, the circulating current in this example is just under 0.16 per unit. This can be verified by the student by solving the loop equations shown in Figure 4.11. Since the tertiary rating is 0.35 per unit (35% of the primary winding MVA rating), this circulating current will not overload the tertiary windings.

4.12 THERMAL CAPABILITY OF THREE-WINDING TRANSFORMERS

Calculation of the thermal capability of a two-winding transformer is straightforward because the magnitude of the KVA in the primary winding equals the KVA in the secondary winding. The calculation of the thermal capability of a three-winding transformer is complicated by the fact that the KVA loads of the three windings may be different and are unpredictable. In fact, a precise calculation of the thermal capability using the methodology in IEEE Standard C57.92-1981 is generally not possible unless certain simplifying assumptions are made. Usually, the tertiary load is either assumed to have a constant load equal to the rating of the tertiary winding or is assumed to have a load equal to the secondary load.

When the primary winding supplies all of the input KVA, it will carry the sum of the secondary and tertiary loads. In this case, the transformer thermal capability is based on the thermal capability of the primary winding as long as neither the secondary load nor tertiary load exceed specified limits. For example, a 100 MVA transformer may have a specified limit of 70 MVA on either the secondary or tertiary, so the transformer can be safely loaded with 70 MVA on the secondary and 30 MVA on the tertiary, with 30 MVA on the secondary and 70 MVA on the tertiary, or with 50 MVA on the secondary and 50 MVA on the tertiary, and so on.

4.13 THE STABILIZING EFFECT OF A Δ TERTIARY WINDING

A Δ-connected tertiary winding is often added to stabilize voltages with Y-Y connections. As an example, consider a Y-Grd.Y transformer with the secondary winding grounded and the primary winding ungrounded. This connection is used to isolate the primary circuit from ground currents in the secondary circuit; however, this connection would ordinarily produce a very bad voltage unbalance when phase-to-neutral loads are connected to the secondary. The root cause for the voltage unbalance is that zero-sequence currents have no path in the primary circuit because the neutral is ungrounded. The secondary zero-sequence currents must then encounter a relatively high magnetizing impedance. Adding a Δ-connected tertiary winding solves the problem by providing grounding bank action. With a local source of ground current, the primary circuit is not required to carry neutral current.

The diagram in Figure 4.12 shows the sequence network connections for a bank of single-phase transformers connected Y-Grd.Y with a 1.0 per

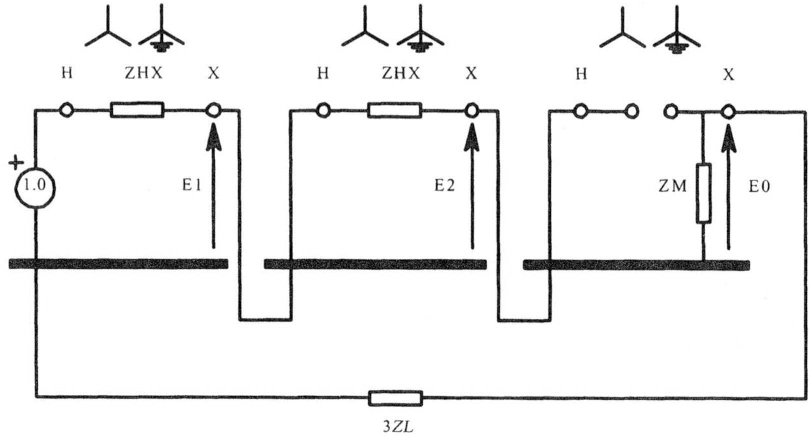

FIGURE 4.12 Positive, negative, and zero phase sequence networks connected to simulate a single-phase load Z_L, connected to the secondary of a Y-Grd.Y transformer. Zero sequence currents must pass through a relatively large magnetizing impedance Z_M, which produces a large zero sequence voltage E_0.

unit single-phase-to-ground load Z_L applied to phase A When the single-phase load is supplied by a Y-Grd.Y transformer bank shown in the figure, the sequence networks show that all of the load current must pass through the magnetizing impedance Z_M, which is typically much greater than 1 per unit. This greatly limits the available load current. The sequence voltages and secondary phase voltages for this connection are summarized below.

$$Z_L \approx 1.0 \tag{4.13.1}$$

$$Z_M \gg 3 \times Z_L \gg Z_{HX} \tag{4.13.2}$$

$$E_1 \approx 1.0 \tag{4.13.3}$$

$$E_2 \approx 0 \tag{4.13.4}$$

$$E_0 \approx -1.0 \tag{4.13.5}$$

$$E_a \approx 0 \tag{4.13.6}$$

$$E_b \approx 1.73\angle 120^\circ \tag{4.13.7}$$

$$E_c \approx 1.73\angle 60^\circ \tag{4.13.8}$$

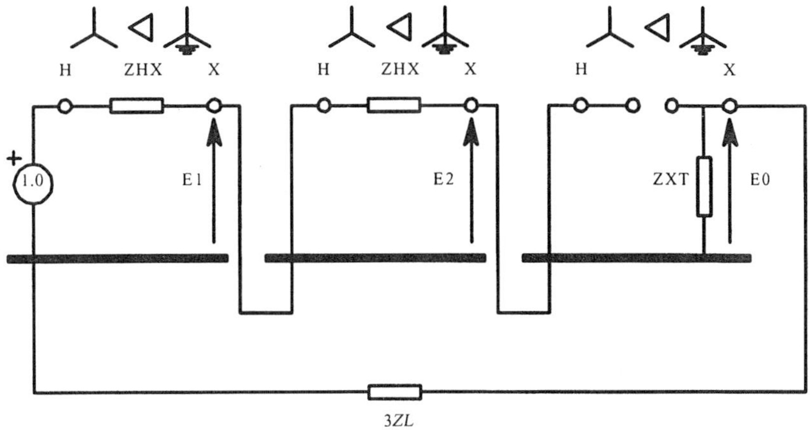

FIGURE 4.13 Positive, negative, and zero phase sequence networks connected to simulate a single-phase load Z_L, connected to the secondary of a Y-Grd.Y transformer with a Δ-connected tertiary winding added. Zero sequence currents pass through a relatively small series impedance Z_{XT} between the secondary and tertiary windings.

The secondary voltage E_a on the loaded phase collapses and the secondary voltages on the other two phases, E_b and E_c, approach the magnitudes of the phase-to-phase voltages. This may drive the core into saturation and reduce Z_M; however, a transformer with a saturated core does not behave as a normal transformer. The actual voltages are difficult if not impossible to determine under these conditions.

The diagram shown in Figure 4.13 shows the sequence network connections for a bank of single-phase transformers with Δ-connected tertiary windings added to create a Y-Δ-Grd.Y transformer connection. A 1.0 per unit single-phase-to-ground load is applied to phase A.

When the single-phase load is connected to a Y-Δ-Grd.Y transformer bank, the sequence networks clearly show that the tertiary windings provide an effective ground path through the low-voltage winding to tertiary-winding impedance Z_{XT}. This magnitude of Z_{XT} is slightly greater than the magnitude of the high-voltage to low-voltage impedance, Z_{HX}. The sequence voltages and phase voltages are summarized below.

$$Z_L \approx 1.0 \tag{4.13.9}$$

$$3 \times Z_L \gg Z_{XT} > Z_{HX} \tag{4.13.10}$$

$$E_1 \approx 1.0 \tag{4.13.11}$$

$$E_2 \approx 0 \tag{4.13.12}$$

$$E_0 \approx 0 \tag{4.13.13}$$

$$E_a \approx 1.0 \tag{4.13.14}$$

$$E_b \approx 1.0\angle 240^\circ \tag{4.13.15}$$

$$E_c \approx 1.0\angle 120^\circ \tag{4.13.16}$$

The secondary voltage E_a on the loaded phase as well as the secondary voltages on the other two phases are maintained at about the same values as when the transformer bank is unloaded. The tertiary winding current is equal to the zero-sequence current, which is one-third of the *A* phase current. Therefore, if the tertiary windings have a KVA rating equal to 35% of the KVA rating of the low-voltage and high-voltage windings, then none of the windings will be overloaded.

One of the common types of large power transformers used in bulk power substations is a three-phase Grd.Y-Grd.Y autotransformer with a Δ-connected tertiary winding. The purpose of the tertiary winding is to stabilize the voltages of the Grd.Y-Grd.Y connection, so its KVA rating is around one-third of the autotransformer's KVA rating. This configuration can be implemented either as a bank of three single-phase transformers or as a three-phase unit. There are several inherent advantages of this design. First of all, using an autotransformer for the main windings takes advantage of the considerable savings in size, weight, and economy of the autotransformer. The Y-Y connection allows systems of various voltages to be tied together without introducing phase shifts between them, and most of the inherent disadvantages of the Y-Y connection are ameliorated by the influence of the Δ-connected tertiary. Finally, by choosing a distribution voltage for the voltage rating of the tertiary winding, the tertiary can supply station service or auxiliary load.

Example 4.7

The sample test report shown in Figure 4.14 is for a typical large three-winding transformer. Using the information shown, determine the *T* equivalent circuit impedance values, expressed in per unit of 84 MVA. The impedance values shown on the nameplate are expressed on different KVA bases. Assuming that the impedances are mostly reactive, the reactive components are calculated as follows.

REPORT OF TEST ON TRANSFORMERS

Date of Test June 11, 1947

Purchaser TransCo Pennsylvania, Macungie, PA

RATING Type FOA 3 Phase 60 Cycles **75000/84000 (55/65C) KVA

	H - Winding	X - Winding	Y - Winding
Volts	230000GRY	69000GRY	12470
Kva	**	**	26250/29400

Taps as per attached Nameplate 115387 - C

CHARACTERISTICS

Losses, efficiencies and regulations are based on wattmeter measurments and, unless otherwise stated, on normal rating. For three phase transformers the resistances given are the sum of the three phases.

Serial	Resistance at 75 C			NO LOAD LOSS WATTS AT 100% VOLTAGE	% EXCITING CURRENT AT 100% VOLTAGE	230GRY Kv To 69GRY Kv At 75000 Kva		230GRY Kv To 12.47 Kv At 26250 Kva		69GRY Kv To 12.47 Kv At 26250 Kva	
	H	X	Y			LOAD LOSS WATTS 75 C	% IMP. 75 C	LOAD LOSS WATTS 75 C	% IMP. 75 C	LOAD LOSS WATTS 75 C	% IMP. 75 C
D572984	5.347	0.2584	0.0648	77270	0.543	338000	15.31	75780	12.02	52230	6.70
AVERAGE											
GUARANTEE	——	——	——	79000	1.0	——	15.50	——	12.75	——	7.00

Total Loss Watts at 75 C		230GRY–69GRY	230GRY–12.47	69GRY–12.47
	AVERAGE	415270	153050	129500
	GUARANTEE	438000		

	EFFICIENCIES AT 75C 100% P-F						REGULATION AT 75C			
	Load	Full Load	3/4 Load	1/2 Load	1/4 Load		100% P-F	90% P-F	80% P-F	70% P-F
AVERAGE		99.45	99.53	99.57	99.48	AVERAGE				
GUARANTEE						GUARANTEE				

TEMPERATURE RISE (Serial No. *)

Average ultimate temperature rise in deg C corrected to instant of shutdown

		H - Winding		X - Winding		Y - Winding		
Winding Connection		230000GRY V.		69000GRY V.		12470 V.		Method of Cooling FOA
MVA	Amp.	84	211	84	703	29.4	1360	Top Oil Rise 43.5 C
Rise: Test	Guar.	55.4 C	65 C	54.0 C	65 C	55.4 C	65 C	Ambient 25 C
Winding Temp. Ind.		deg C		deg C		deg C		
Winding Connection		V.		V.		V.		Method of Cooling
MVA	Amp.							Top Oil Rise C
Rise: Test	Guar.	C	C	C	C	C	C	Ambient C
Winding Temp. Ind.		deg C		deg C		deg C		

FIGURE 4.14 A typical test report for a three-winding transformer.

$$X_{HX} = 0.1531 \times \frac{84 \text{ MVA}}{75 \text{ MVA}} = 0.1715 \text{ per unit}$$

$$X_{HY} = 0.1202 \times \frac{84 \text{ MVA}}{26.25 \text{ MVA}} = 0.3846 \text{ per unit}$$

$$X_{XY} = 0.0670 \times \frac{84 \text{ MVA}}{26.25 \text{ MVA}} = 0.1920 \text{ per unit}$$

The reactance values between windings are converted to winding reactances using Equations 4.10.1 through 4.10.3:

$$X_H = \frac{0.1715 + 0.3846 - 0.1920}{2} = 0.1821 \text{ per unit}$$

$$X_X = \frac{0.1715 + 0.1920 - 0.3846}{2} = -0.02121 \text{ per unit}$$

$$X_Y = \frac{0.3846 + 0.1920 - 0.1715}{2} = 0.2026 \text{ per unit}$$

The resistance values shown on the test report are the sum of the three phases. These are converted to per unit values per phase at the nominal voltage rating.

$$R_H = 5.347\ \Omega \times \frac{84{,}000{,}000 \text{ VA}}{3 \times (230{,}000V)^2} = 0.00283 \text{ per unit}$$

$$R_X = 0.2584\ \Omega \times \frac{84{,}000{,}000 \text{ VA}}{3 \times (69{,}000V)^2} = 0.00152 \text{ per unit}$$

$$R_Y = 0.06479\ \Omega \times \frac{84{,}000{,}000 \text{ VA}}{3 \times (12{,}470V)^2} = 0.0117 \text{ per unit}$$

$$Z_H = R_H + jX_H = 0.00283 + j0.1821$$

$$Z_X = R_X + jX_X = 0.00152 - j0.02121$$

$$Z_Y = R_Y + jX_Y = 0.0117 + j0.2026$$

REFERENCE

1. R.L. Bean, N. Chackan, Jr., H. R. Moore, and E. C. Wentz. *Transformers for the Electric Power Industry*. McGraw-Hill, New York, 1959, p. 236.

5

Short Circuits, Inrush Currents, and Other Phenomena

5.1 EFFECTS OF SHORT CIRCUITS ON TRANSFORMERS

Transformers are susceptible to damage by secondary short-circuit currents having magnitudes that can be many times rated load current. The damage results from the following effects:

- The I^2R losses in the winding conductors are increased by the square of the current. This increases the temperature rise of the windings. Because protective devices limit the duration of short circuits (as opposed to overloads), the temperature rise of the winding can be calculated by dividing the total energy released by the I^2R losses by the thermal capacity of the conductor.
- The short-circuit currents exclude flux in the core and increase stray flux around the core. This stray flux induces currents in metallic parts other than the winding conductors, which can be damaged thermally.
- As we saw in Chapter 4, a short circuit applied to the secondary circuit of an autotransformer can substantially increase the voltage across the series winding and across the common winding through

induction. This not only presents the possibility of damaging the winding insulation by overvoltage, but will also drive the core into saturation and significantly increase core losses with potential damaging effects from temperature.

- Bushings and tap changers have current ratings that are usually only marginally greater than the rated load of the transformer. Since fault currents are many times rated currents and these components have short thermal time constants, they can be seriously overloaded and thermally damaged.
- Stray flux in the vicinity of current-carrying conductors produces mechanical forces on the conductors. When a short circuit is applied to a transformer, there is a significant increase in stray flux, resulting in greater mechanical forces on the windings, leads, bushings, and all other current-carrying components. These components, especially the windings, must be braced to withstand these forces.

A good transformer design must take all of the above effects into account to minimize the risk of damage and assure a long service life.

5.2 COMPARISONS OF SHORT-CIRCUIT CURRENTS FOR VARIOUS FAULTS

Since we know (at least approximately) the equivalent circuit of a transformer by calculating the impedance values from factory tests, it is relatively straightforward to calculate the short-circuit currents resulting from various types of short circuits or *faults*. There are several types of faults that occur in transmission and distribution systems, including phase-to-ground faults, phase-to-phase faults, phase-to-phase-to-ground faults, three-phase grounded faults and three-phase ungrounded faults. One type of fault is usually more severe than the others in terms of generating damaging currents. This is called the *limiting fault*, because fault current can go no higher for any other fault. The limiting fault depends on the transformer connection and system impedances and is not always obvious.

The general short-circuit requirements for liquid-immersed distribution, power, and regulating transformers are specified in Section 7 of IEEE Std. C57.12.00-2000 [1, pp. 35–38]. The duration of the short-circuit current that the transformer must withstand is given by the following formula:

$$t = \frac{1250}{I^2} \tag{5.2.1}$$

where

t = duration of the fault, s
I = symmetrical short-circuit current in per unit

For transformers rated above 500 KVA, the standard limits the duration of the fault to 2 s. The magnitudes of fault currents are calculated according to the rules given in Table 5.1.

A transformer is required to withstand a first-cycle asymmetrical peak current, I_{sc}(pk asym), calculated as follows.

$$\begin{aligned} I_{sc}(\text{pk asym}) &= K \times I_{sc} = 1.414 \\ &\quad \times (1 + e^{-(\phi+\pi/2)r/x} \times \sin\phi) \times I_{sc} \end{aligned} \tag{5.5.2}$$

where

I_{sc} = symmetrical fault current, RMS
$\phi = \tan^{-1}(x/r)$
x/r = effective ratio of the AC reactance to resistance of the total impedance that limits the fault current for the transformer connections used

TABLE 5.1 Accepted Methods for Calculating Fault Currents by Transformer Category

Category	Single-phase KVA	Three-phase KVA	Fault current calculation
I	5–25	15–75	Transformer impedance only, limited to 40 per unit current
I	37.5–110	112.5–300	Transformer impedance only, limited to 35 per unit current
I	167–500	500	Transformer impedance only, limited to 25 per unit current
II	501–1667	501–5000	Transformer impedance only, no fault current limit
III	1,668–10,000	5001 30,000	Transformer impedance plus system impedance
IV	Above 10,000	Above 30,000	Transformer impedance plus system impedance

Eq. (5.2.2) accounts for the initial *DC offset* and its subsequent decay when the peak of the AC wave occurs. The initial DC offset actually varies, depending on the precise moment when the circuit is short-circuited. The peak value is a statistical function; however, the limiting case is when the AC wave is completely offset initially, as reflected in Eq. (5.2.2). If the x/r ratio is infinite, the DC offset does not decay, and the first peak is equal to 2 times the square root of 2 times the RMS value of the symmetrical fault current.

IEEE Std. C57.12.00-2000 also specifies the methods used for short-circuit testing. Not every transformer that is manufactured is routinely tested in the factory for short-circuit capability; however, a representative sample of every generic transformer *design* is tested to demonstrate the adequacy of the design.

Example 5.1

A three-winding Grd.Y-Grd.Y-Δ transformer has a T equivalent circuit with following impedance values:

$$Z_H = j0.1821 \text{ per unit}$$
$$Z_X = -j0.02121 \text{ per unit}$$
$$Z_Y = j0.2026 \text{ per unit}$$

Calculate the short-circuit currents for each of the following faults at the low-voltage with the high-voltage winding connected to an infinite bus (a system with zero impedance):

1. Three-phase ungrounded fault
2. Phase-to-ground fault
3. Phase-to-phase-to-ground fault
4. Phase-to-phase fault

The positive-sequence, negative-sequence, and zero-sequence network connections for the three-phase ungrounded fault are shown in Figure 5.1. In short-circuit calculations the positive-sequence system voltage source is usually assumed to be 1.0 per unit with no negative-sequence or zero-sequence voltage sources. Thus, only the positive-sequence network is involved in a three-phase fault. Note that the series impedance of the transformer in the positive-sequence network Z_{HX} is simply the sum of Z_H and Z_X.

The positive-sequence fault current is equal to the voltage divided by the series impedance: $I_1 = 1.0/j0.1609 = -j6.215$ per unit.

Equations (3.6.4) to (3.6.6) are used to convert the sequence currents into phase currents:

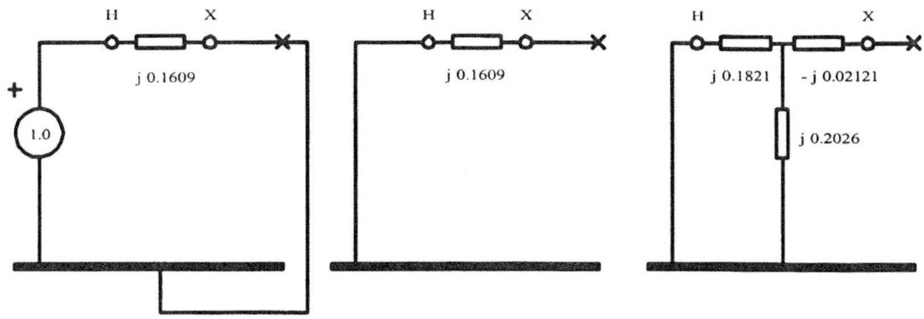

FIGURE 5.1 Positive, negative, and zero phase sequence networks connected for a three-phase fault.

$$I_A = I_0 + I_1 + I_2 = 0 - j6.215 + 0 = 6.215\angle 270^\circ$$
$$I_B = I_0 + I_1\angle 240^\circ + I_2\angle 120^\circ = 0 - j6.215\angle 240^\circ + 0 = 6.215\angle 150^\circ$$
$$I_C = I_0 + I_1\angle 120^\circ + I_2\angle 240^\circ = 0 - j6.215\angle 120^\circ + 0 = 6.215\angle 30^\circ$$

The positive-sequence, negative-sequence, and zero-sequence network connections for the phase-to-ground fault are shown in Figure 5.2. Since the system impedance is zero, the high-voltage terminals are short-circuited in the negative-sequence and zero-sequence networks. In the zero-sequence network the impedance from H to X is the parallel combination of Z_H and Z_Y in series with Z_X:

$$Z_{HX} = Z_X + \left[\frac{1}{Z_H} + \frac{1}{Z_Y}\right]^{-1} = -j0.02121 + [-j5.4915 - j4.9358]^{-1}$$
$$= j0.07469 \text{ p.u.}$$

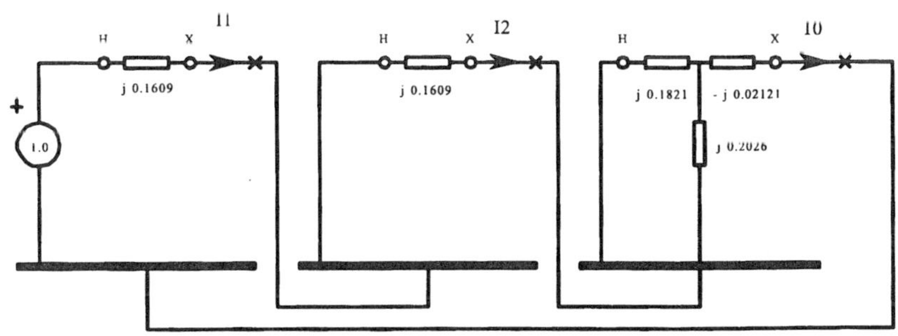

FIGURE 5.2 Positive, negative, and zero phase sequence networks connected for a phase-to-ground fault.

The currents at the point of the fault in the positive-sequence, negative-sequence, and zero-sequence networks are equal:

$$I_1 = I_2 = I_0 = \frac{1.0}{j0.1609 + j0.1609 + j0.07469} = -j2.5221 \text{ per unit}$$

Equations (3.6.4) to (3.6.6) are used to convert the sequence currents into phase currents:

$$I_A = I_0 + I_1 + I_2 = -j2.5221 - j2.5221 - j2.5221$$
$$= 7.5663\angle 270°$$
$$I_B = I_0 + I_1\angle 240° + I_2\angle 120°$$
$$= -j2.5221 - j2.5221\angle 240° - j2.5221\angle 120° = 0$$
$$I_C = I_0 + I_1\angle 123° + I_2\angle 240°$$
$$= -j2.5221 - j2.5221\angle 120° - j2.5221\angle 240° = 0$$

The positive-sequence, negative-sequence, and zero-sequence network connections for the phase-to-phase-to-ground fault (*B* to *C* to ground) are shown in Figure 5.3. As in the phase-to-ground fault, the impedance from *H* to *X* in the zero-sequence network is equal to j0.07469 per unit. The total impedance in series with the positive-sequence voltage source Z_S is calculated below.

$$Z_S = j0.1609 + \left[\frac{1}{j0.1609} + \frac{1}{j0.07469}\right]^{-1} = j0.2119 \text{ per unit}$$

The positive-sequence current at the point of the fault is $I_1 = 1.0/Z_S = -j4.7190$ per unit. This current is split into parallel negative-sequence and

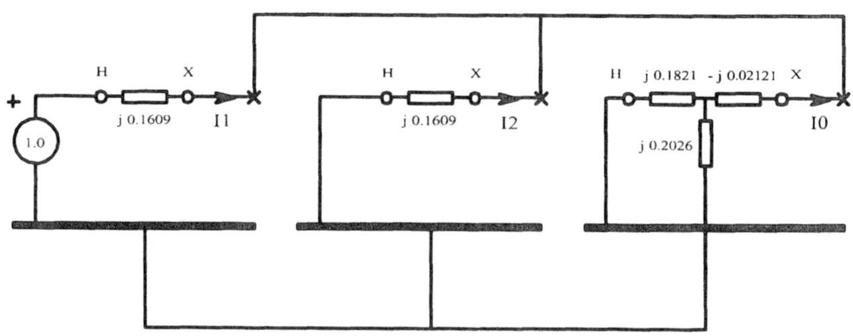

FIGURE 5.3 Positive, negative, and zero phase sequence networks connected for a phase-to-phase-to-ground fault.

zero-sequence currents at the point of the fault. Taking the current flowing *out* of the X terminal as *positive* current:

$$I_2 = -I_1 = \frac{1/\text{j}0.1609}{1/\text{j}0.1609 + 1/\text{j}0.07469} = -I_1 \times 0.3170$$

$$= \text{j}1.4961 \text{ per unit}$$

$$I_0 = -I_1 - I_2 = \text{j}3.2229 \text{ per unit}$$

Equations (3.6.4) to (3.6.6) are used to convert the sequence currents into phase currents:

$$I_A = I_0 + I_1 + I_2 = \text{j}3.2229 - \text{j}4.7190 + \text{j}1.4961 = 0$$

$$I_B = I_0 + I_1\angle 240° + I_2\angle 120° = \text{j}3.2229 - \text{j}4.7190\angle 240° + \text{j}1.4961\angle 120°$$

$$= 7.2347\angle 138.1°$$

$$I_C = I_0 + I_1\angle 120° + I_2\angle 240° = \text{j}3.2229 - \text{j}4.7190\angle 120° + \text{j}1.4961\angle 240°$$

$$= 7.2347\angle -41.9°$$

The positive-sequence, negative-sequence, and zero-sequence network connections for the phase-to-phase fault (B to C) are shown in Figure 5.4. Only the positive-sequence and negative-sequence networks are involved in a phase-to-phase fault; therefore, $I_0 = 0$. The total impedance in series with the positive-sequence voltage source is the sum of the transformer series impedance values of the positive-sequence and negative-sequence networks:

$$Z_S = \text{j}0.1609 + \text{j}0.1609 = \text{j}0.3218 \text{ per unit}$$

$$I_1 = -I_2 = \frac{1.0}{\text{j}0.3218} = -\text{j}3.1075 \text{ per unit}$$

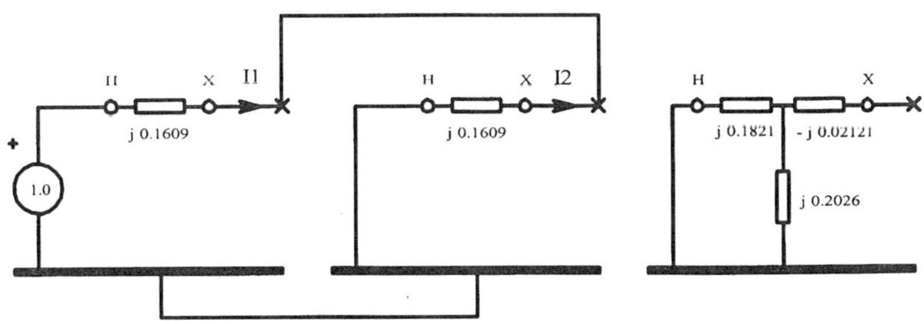

FIGURE 5.4 Positive, negative, and zero phase sequence networks connected for a phase-to-phase fault.

TABLE 5.2 Sequence Currents and Phase Currents for Various Types of Faults Applied to a Three-Winding Transformer in Example 5.1

Type of fault	I_1	I_2	I_0	I_A	I_B	I_C
Three phase	6.215	0	0	6.215	6.215	6.215
A phase to ground	2.522	2.522	2.522	7.566	0	0
B phase to *C* phase to ground	4.719	−1.496	−3.223	0	7.235	7.235
B phase to *C* phase	3.108	−3.108	0	0	5.382	5.382

Equations (3.6.4) to (3.6.6) are used to convert the sequence currents into phase currents:

$$I_A = I_0 + I_1 + I_2 = 0 - j3.1075 + j3.1075 = 0$$
$$I_B = I_0 + I_1\angle 240° + I_2\angle 120° = 0 - j3.1075\angle 240° + j3.1075\ \angle 120°$$
$$= 5.3823\angle 180°$$
$$I_C = I_0 + I_1\angle 240° + I_2\angle 120° = 0 - j3.1075\angle 120° + j3.1075\ \angle 240°$$
$$= 5.3823\angle 0°$$

The sequence currents values and the *magnitudes* of the phase currents at the low-voltage transformer terminals are tabulated in Table 5.2. (The phase angles of I_1, I_2, and I_0 are equal to $-90°$.) The limiting fault for this transformer is the phase-to-ground fault, because it has the largest magnitude of fault current = 7.566 per unit.

Example 5.2

Assume the transformer in Example 5.1 has a 150 MVA rating, the low-voltage winding is rated 69 kV, and that all of the transformer impedance elements have an x/r ratio of 20 : 1. Calculate the first cycle asymmetrical peak fault current for the limiting fault.

Using Eq. (5.2.2),

$$I_{sc}(\text{pk asym}) = K \times I_{sc}$$

$$\phi = \tan^{-1}\frac{x}{r} = \tan^{-1} 20 = 1.52 \text{ rad}$$

$$\sin\phi = \sin 1.52 = 0.999$$

$$K = 1.414 \times (1 + e^{-(1.52+\pi/2)/20} \times 0.999\ ) = 2.625$$

The limiting fault in Example 5.1 is a phase-to-ground fault. This fault has a magnitude of 7.566 per unit.

$$I_{sc}(\text{pk asym}) = 2.625 \times 7.566 = 19.86 \text{ per unit}$$

The current base at the low-voltage terminals is calculated below:

$$I_{base} = \frac{\text{KVA}_{base}}{1.732 \times \text{kV}_{base}} = \frac{150{,}000 \text{ KVA}}{1.732 \times 69 \text{ kV}} = 1255 \text{ A}$$

Converting I_{sc} from per unit to amperes,

$$I_{sc}(\text{pk asym}) = 19.86 \times 1255 \text{ A} = 24{,}926 \text{ A}$$

5.3 MECHANICAL FORCES IN TRANSFORMERS

While the duration of fault current is not normally long enough to cause thermal damage, the damaging effects of mechanical forces become a very important design consideration for large power transformers. For a two-winding transformer carrying normal load current, the current in the primary winding flows in a direction opposite to the direction of the current in the secondary winding, and the total ampere-turns in the primary are equal to and opposite from the total ampere-turns in the secondary. Because there is mutual stray flux between the windings, the primary and secondary windings tend to repel each other. The repulsive forces are greatly magnified for the transformer under fault conditions.

The way to approach the analysis of forces on transformer windings is to consider only the forces on a conductor caused by leakage flux passing *through* the conductor. Note that the flux in the core does not exert any force on the conductor because none of the flux in the core will pass through the conductor. There is a well-known formula in physics for calculating the force on a conductor. Sticking to the MKS system of units:

$$\mathbf{F} = \mathbf{i} \times \mathbf{B} \tag{5.3.1}$$

where

$\mathbf{F}$ = force vector on the conductor, N/m
$\mathbf{i}$ = current vector through the conductor, A
$\mathbf{B}$ = flux density vector through the conductor, Wb/m^2

The three quantities in the formula are vectors and $\mathbf{i} \times \mathbf{B}$ is the cross product of $\mathbf{i}$ times $\mathbf{B}$. When applying the cross product, the force vector will have a component in each of the three directions x, y, and z. The current and the flux density also have three components. The relationships between the components of force and components of current and flux density resulting from the cross product **i** × **B** are given in the equations below:

$$F_x = i_y B_z - i_z B_y \tag{5.3.2}$$

$$F_y = i_z B_x - i_x B_z \tag{5.3.3}$$

$$F_z = i_x B_y - i_y B_x \tag{5.3.4}$$

In Chapter 1, we learned that the magnetic flux density surrounding a long, straight conductor is directly proportional to the conductor current and inversely proportional to the distance from the center of the conductor. If two long parallel conductors carry currents in the same direction, the directions of the magnetic fields with respect to the currents will produce forces that are equal and opposite, tending to pull the conductors together. If two long parallel conductors carry currents in opposite directions, the directions of the magnetic fields with respect to the currents will produce forces that are equal and opposite, tending to push the conductors apart. This is illustrated in Figure 5.5.

The forces must be equal and opposite because there can be no net force produced on the pair of conductors; otherwise, there would be a "levitating force" present. As Figure 5.5 illustrates, the only component of current is in the $\pm z$ direction (normal to the plane of the page) and the only component of flux at the centers of each conductor is in the $\pm y$ direction (vertical). Therefore, the only component of force is in the $\pm x$ direction (horizontal). Therefore, Eqs. (5.3.3) and (5.3.4) each produces a zero force, and

$$F_x = -i_z B_y \tag{5.3.5}$$

The force in each conductor is equal to the current in that conductor times the flux density produced by the *other* conductor. This flux density is equal to μ_0 times the current in the other conductor divided by 2π times the distance r between the conductors.

$$|F_x| = i_1 \times i_2 \times \frac{\mu_0}{2\pi r} \tag{5.3.6}$$

Equation (5.3.6) confirms that the forces on the two conductors must have equal magnitudes, even if the currents have unequal magnitudes. As stated earlier, these forces are *always* in opposite directions; however, the forces can either be repulsive or attractive, depending on the relative directions of the currents.

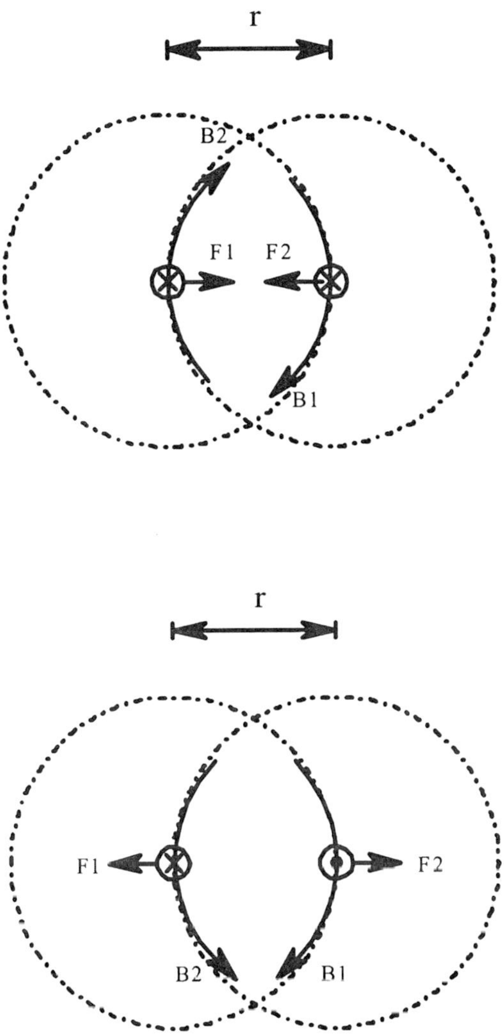

FIGURE 5.5 Forces acting on a pair of straight current-carrying conductors. F1 and F2 are always equal and opposite.

Example 5.3

Calculate the force between two straight conductors with centers 3 cm apart, carrying 10,000 A (RMS) in opposite directions. The peaks of the mutual forces are based on the *peak* values of the currents, which are 1.414 times the RMS current values.

$$|F_x| = (1.414 \times 10{,}000 \text{ A})^2 \times \frac{4\pi \times 10^{-7} \text{ N/A}}{2\pi \times 0.03 \text{ m}}$$

$$= 1333 \text{ N/m} = 914 \text{ lb/ft}$$

Since the currents are in opposite directions, the mutual forces will be repulsive.

5.4 FORCES BETWEEN TRANSFORMER WINDINGS

Example 5.3 illustrates that tremendous forces exist between one pair of conductors that are spaced fairly close together and carry currents that are comparable to the short-circuit currents encountered in power systems. In transformer windings, there are many conductors spaced closely together, and when faults occur the forces acting on these conductors can be truly mind-boggling. With mutual forces existing between *every pair* of conductors, calculating all of these mutual forces on even *one* conductor is a very formidable task. This task is complicated by the fact that the geometry of the windings is not even close to the simple long and straight pairs of conductors. Adding even more complication is the fact that the leakage flux is distorted by the presence of the iron core, which also has a complicated geometry, so the most difficult part is determining the magnitudes and directions of the leakage flux produced by all these currents.

Calculating short-circuit forces in a transformer using analytical methods is, in fact, impossible. Some fairly accurate quantitative analysis is possible with the use of modern computer-aided design tools, such as finite-element modeling.

Due to the difficulty in computing short-circuit forces accurately, transformers in the early years of transformer technology were built with plenty of mechanical "design margin" which was determined mostly by trial and error. In the early 1970s when the first primitive computer-aided design tools came into existence, many transformer manufacturers attempted to remove some of the inherent ''fat'' of the earlier transformer designs in order to gain an economic advantage over their competitors. Unfortunately, some manufacturers cut into the bone as well, and transformers of the 1970s vintage have a notorious reputation for lacking short-circuit capability. By the end of the twentieth century, most transformers of that vintage had permanently removed themselves from service.

Doing any sort of quantitative analysis of short-circuit forces is a task for the transformer design engineer and is obviously beyond the scope of this book. But in order to gain at least a fundamental understanding and appreciation of short-circuit forces without getting bogged down in the complexity of the math, a qualitative approach can be used instead.

Consider a two-winding core-form transformer. Under short-circuit conditions, the total ampere turns in the primary winding are (approximately) equal and opposite from the total ampere-turns in the secondary winding. The magnitudes of the primary and secondary ampere turns are only approximately equal because of the presence of the magnetizing impedance. This creates a shunt path that subtracts ampere-turns from one of the windings. Neglecting this effect, both windings can be modeled as having the same number of turns and carrying the same current, with the currents located at the *magnetic centers* of the coils, as illustrated in Figure 5.6. The three-dimensional figure on the left-hand side of Figure 5.6 shows the primary and secondary windings as concentric cylinders with currents flowing in opposite directions along the magnetic centers. The cutaway view on the right-hand side of Figure 5.6 depicts the magnetic forces operating at the magnetic centers. Modeling the windings as solid composite structures instead of groups of individual conduc-

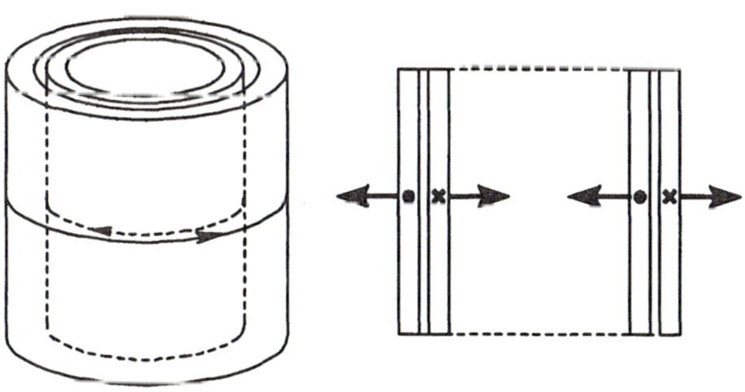

FIGURE 5.6 A pair of cylindrical concentric windings with their magnetic centers perfectly aligned.

tors greatly simplifies the problem and aids visualization of the magnetic forces involved.

The radial forces acting on the primary and secondary coils tend to crush the inner coil (usually the secondary) toward the center of the core and expand the outer coil (usually the primary) outward. If the coils have a rectangular cross-sectional design instead of a circular cross section design, then the outer coil tends to become circular under short-circuit conditions, thus distorting its shape. The outward radial forces result in hoop stresses in the conductors. The tensile strength of the outer conductors is generally one of the major limitations in the short-circuit capability of the outer winding due to these hoop stresses. The inner winding encounters compressive forces that are dealt with by bracing the inner winding against the core.

Unless the magnetic centers of the primary and secondary windings are aligned perfectly, which is rarely the case, the magnetic forces that repel the windings will have vertical components as well as radial forces. This situation is illustrated in Figure 5.7.

Since most transformers have tapped primary windings, it is difficult to align the magnetic centers of the primary and secondary windings for every tap position. For tap positions other than the full-tap position, a portion of the primary winding carries no current, creating a "hole" in the winding and moving the magnetic center away from the hole. One remedy for this is to bring out the taps in two groups, one group above and the other group below the center of the winding so that two "holes" are positioned at equal distances from the center of the winding. This aligns the magnetic center with the geometric center of the winding on every tap position, but this also requires two

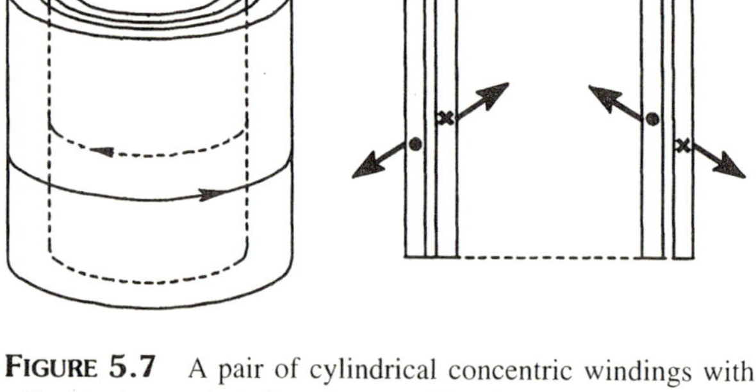

FIGURE 5.7 A pair of cylindrical concentric windings with their magnetic centers offset in the vertical direction, resulting in net vertical forces on the windings.

tap-changing mechanisms per phase. The increased cost of the additional tap-changing equipment may be offset by other cost savings that result from reducing short-circuit forces, so there is a design trade-off in this regard.

Each conductor is pulled toward the magnetic center of its own winding and repelled from the magnetic center of the opposite winding. Because the conductor has a different proximity to each of the two magnetic centers, there will be a net vertical force that tends to pull the conductors toward the middle of the windings.

In a disc-type winding, the conductors are wound in a spiral pattern that forms a series of flat discs. Such a disc-type winding is shown in Figure 5.8 with the winding discs separated from each other by means of radial spacers that are aligned in vertical columns. Under short-circuit forces, the discs are compressed together creating vertical beam stresses on conductors spans between the spacers. The distances between the radial spacers increase away from the center of the winding and if these distances are too great, the vertical

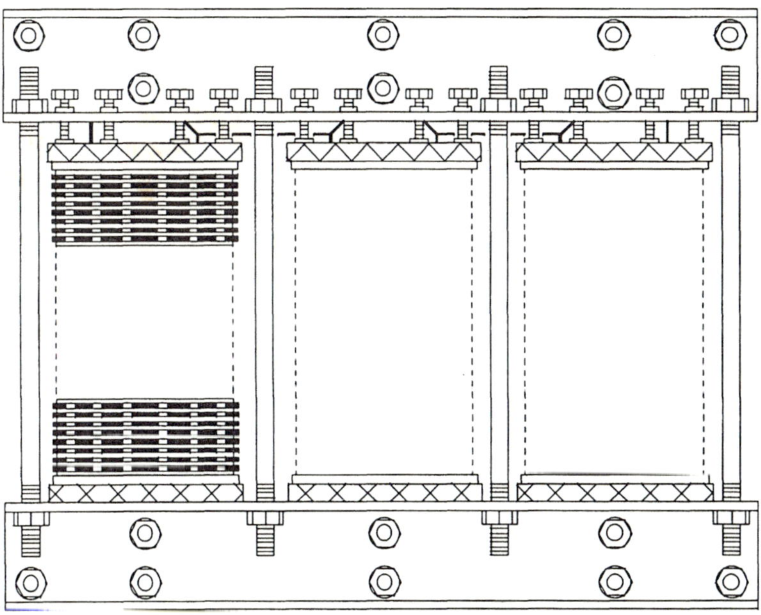

FIGURE 5.8 A three-phase core and coil assembly showing the clamping structure consisting of pressure plates, jack bolts, top and bottom channels, and tie rods.

beam stresses may distort the winding. Therefore, the number of radial spacers is another critical design feature.

In order to control the vertical short-circuit forces, disc windings in core form transformers are kept under tremendous compressive forces by means of various coil clamping systems. The compressive clamping forces keep the windings stationary and under compression at all times in order to prevent damage from distortion. One such coil clamping system for a three-legged core form transformer is depicted in Figure 5.8.

Referring to Figure 5.8, each set of windings on each core leg is compressed between two thick pressure plates, shown as the cross-hatched components in the figure. The pressure plates are made of a strong material that also has good electrical insulating properties. Specially treated laminated wood or epoxy-based materials are suitable for these pressure plates. In the design shown in Figure 5.8, the pressure plates bear against top and bottom C channels that are connected by eight tie rods. Rings of jack bolts press down on the top pressure plates, compressing the windings and tending to force the top and bottom C channels apart. The eight tie rods restrain the forces between the top and bottom C channels. The total force applied through the jack bolts and restrained by the tie rods is greater than the maximum net vertical short-circuit forces that are calculated (or estimated) on the windings.

In a shell-form transformer, radial forces tend to compress the pancake coils toward their magnetic centers instead of exploding the coils outward, unlike the high-voltage coil in a typical core-form transformer. Compression forces are usually easier to deal with than tensile forces, which gives the shell-form transformer a distinct advantage over the core-form design. Since the high-voltage coils are normally sandwiched between the low-voltage coils in a shell-form transformer, the axial forces tend to compress the high-voltage pancake coils toward the center of the winding pack and force the low-voltage pancake coils away from the center of the winding pack. The axial forces are restrained by fill material that is wedged between the coils, between the winding packs, and between the low-voltage coils on the outer winding packs and the transformer tank, which provides mechanical bracing. The resulting configuration, shown in Figure 5.9, is an assembly that is securely braced in all directions and has great mechanical short-circuit strength.

The fill material is shown in Figure 5.9 by the cross-hatched areas. These filled areas are not actually solid, but contain numerous ducts for cooling oil to flow. The transformer tank is divided into an upper and lower half. Wedges at the end surfaces of the windings compress the entire coil structure as the upper tank is lowered on top of the lower tank, resulting in a structure that resists all movement.

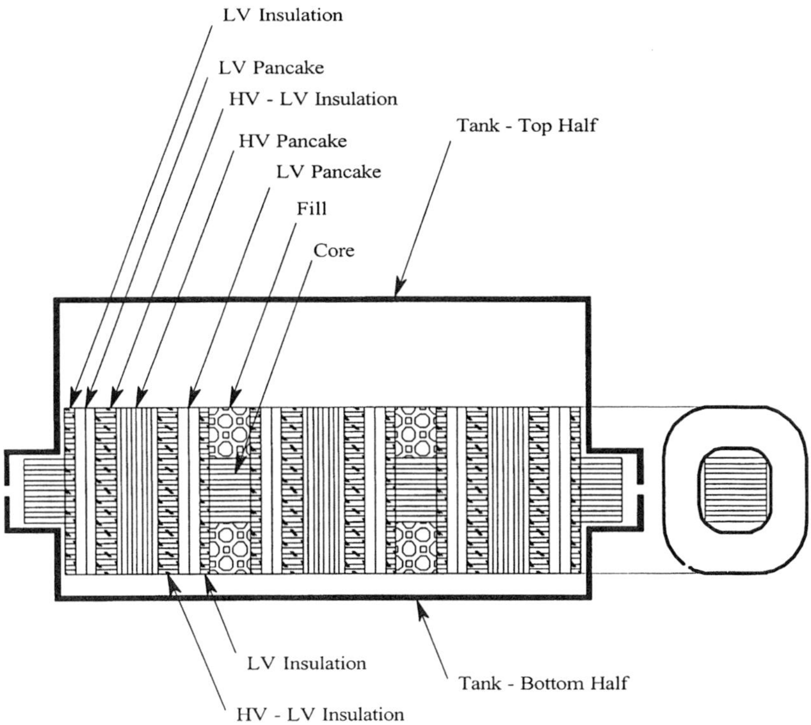

FIGURE 5.9 A three-phase shell-form transformer cutaway sketch. The pancake coil packs are compressed between the ends of the transformer tank, forming a solid bulk.

5.5 SHORT-CIRCUIT FORCES IN THREE-WINDING TRANSFORMERS

The addition of a tertiary winding complicates the already complicated calculation of short-circuit forces because the ampere-turns of the three windings are not necessarily equal. Depending on the type of fault, the currents in the primary, secondary, and tertiary windings can be different in both magnitude and direction. Consider the case of a Y-Grd.Y-Δ transformer with a phase-to-ground fault applied to the secondary winding. Since the primary winding is ungrounded, all of the zero-sequence current flows in the tertiary winding as a closed circulating current. This condition is illustrated as a vector diagram shown in Figure 5.10.

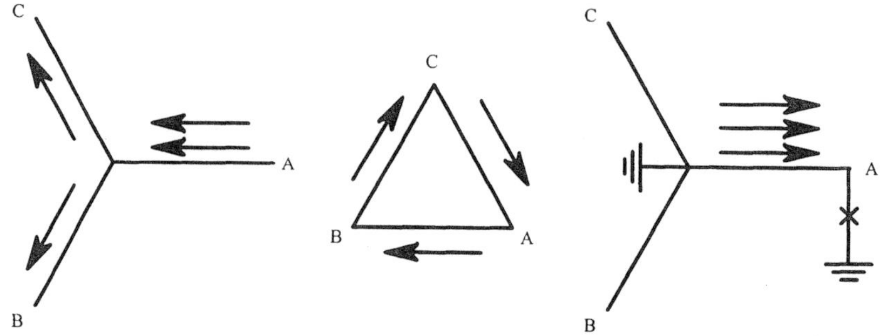

FIGURE 5.10 A phase-to-ground fault applied to the secondary of a three-winding transformer. Each arrow represents 1 unit of ampere-turns.

It is assumed that the total ampere-turns is approximately zero in all cases. Each current arrow in Figure 5.10 represents one "unit" of ampere-turns for the phase-to-ground fault applied to the secondary winding. Knowing the actual numbers of turns or the actual currents in each of the windings is not important in this analysis. In the A phase leg of the transformer, the primary winding has +2 units, the secondary winding has −3 units, and the tertiary winding has +1 unit. In the *B* phase and *C* phase legs, the primary windings have −1 unit each, the secondary windings have 0 units, and the tertiary windings have +1 unit each.

Assume that the physical three-winding transformer has a core-form design with the windings ordered from the center of the core as follows: tertiary, secondary, and primary. The ampere-turns for a phase-to-ground fault applied to the secondary are illustrated in Figure 5.11.

The primary (outer), secondary (middle), and tertiary (inner) windings and their magnetic centers are shown schematically in Figure 5.11. Each symbol "•" represents one unit of ampere-turns flowing out of the page and each symbol "×" represents one unit of ampere-turns flowing into the page. The directions of the net radial forces can be ascertained from the directions of the ampere-turns as shown. These forces tend to expand the primary winding outward and collapse the secondary and tertiary windings inward toward the center of the core. From the simple illustration in Figure 5.11, it would appear that the radial forces between the three pairs of windings partially cancel each other out. Thus, the radial forces in a three-winding transformer on the faulted phase are in general somewhat smaller than the radial forces on the faulted phase in a two-winding transformer.

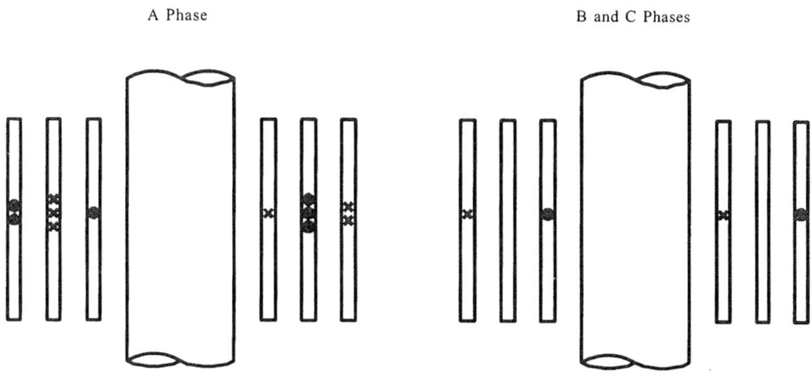

FIGURE 5.11 Distribution of ampere-turns in a three-winding transformer with a phase-to-ground fault applied to the secondary.

5.6 EXCITING CURRENT INRUSH

When a transformer is initially energized, there is a phenomenon known as *exciting current inrush*. Although inrush currents are not generally as damaging as fault currents, the duration of exciting current inrush is on the order of seconds (as compared to on the order of cycles with fault currents). Exciting current inrush conditions also occur much more frequently than short circuits, so this phenomenon is worth exploring.

Consider what happens when initially energizing a single-phase transformer. From Chapter 1 we learned that the flux in the core is equal to the integral of the excitation voltage. If the circuit is closed when the voltage is passing through zero and the initial flux is zero, the sinusoidal flux will be *fully offset* from zero. The full-offset flux has a peak value that is twice the peak value of a symmetrical sinusoidal flux. In other words, the peak flux for a fully offset wave can approach two times the normal peak flux, and this is generally sufficient to drive the core into saturation. At this point, the only thing that limits exciting current is the air-core impedance of the winding, which is several orders of magnitude smaller than the normal magnetizing impedance.

Therefore, the exciting current is much greater than the normal exciting current during the half cycle when the core is saturated. During the opposite half cycle, the core is no longer saturated and the exciting current is approximately equal to the normal exciting current. This is illustrated in Figure 5.12.

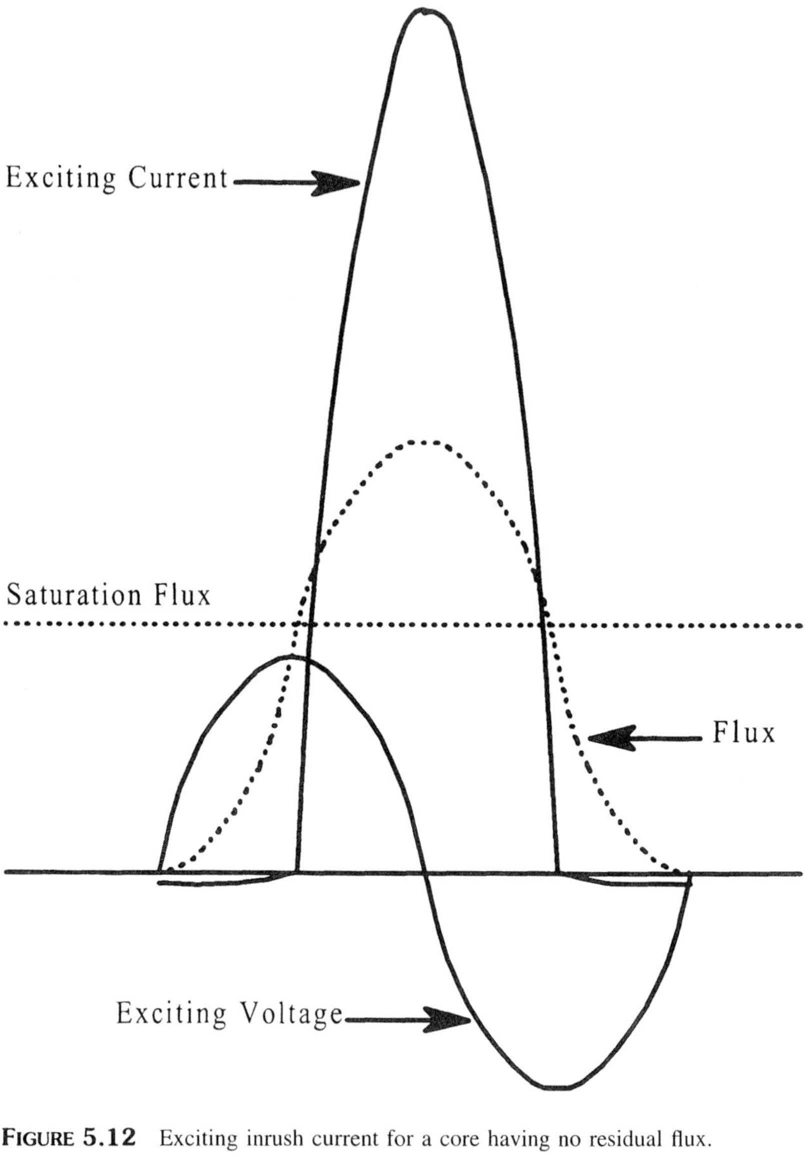

FIGURE 5.12 Exciting inrush current for a core having no residual flux.

The situation is even more extreme when there is *residual flux* in the core and the direction of the residual flux is in the same direction as the offset in the sinusoidal flux wave. This is illustrated in Figure 5.13. Note that Figure 5.12 and 5.13 are drawn on current different scales, so the peak current plotted in Figure 5.13 is actually much larger than the peak current plotted in Figure 5.12.

To find the peak inrush current, limited only by the air-core reactance, it is convenient to calculate the inductance of the winding using cgs units:

$$L = \frac{0.4\pi N^2 A_{\mathrm{mt}}}{l} \tag{5.6.1}$$

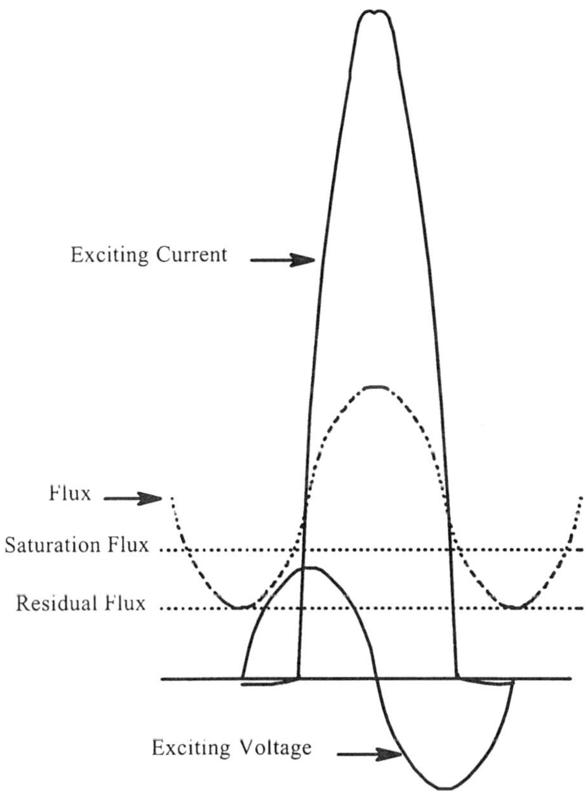

FIGURE 5.13 Exciting inrush current for a core having a residual flux.

where

N = number of turns in the coil
A_{mt} = area inside the mean diameter of the coil, cm^2
l = axial length of the coil, cm
L = inductance of the coil, μH

The flux generated by the inductance ϕ_L is equal to the residual flux plus 2 times the normal flux change minus the saturation flux, since the saturation flux is in the iron. But ϕ_L is related to the inductance and the current:

$$L = \frac{N\phi_L}{I} \tag{5.6.2}$$

Therefore, the peak inrush current is expressed in the cgs system of units as follows [2]:

$$I_{\text{peak}} = \frac{(\phi_r + 2\phi_n - \phi_s)l \times 10^{-8}}{0.4\pi\, NA_{\text{mt}}} \tag{5.6.3}$$

where I_{peak} is in amps and

ϕ_r = residual flux
ϕ_n = normal flux change
ϕ_s = saturation flux

Without resistance in the circuit, each successive peak would have the same value and the current inrush would go on indefinitely. With resistance in the circuit, however, there is a significant voltage drop across the resistance and the flux does not have to rise quite as high as the previous cycle. The integral of the voltage drop represents a net decrease in the flux required to support the applied voltage. Since the $i \times R$ voltage drop is always in the same direction, each cycle decreases the amount of flux required. When the peak value of flux falls below the saturation value of the core, the inrush current disappears. The rate of decay is not exponential although it resembles an exponentially decaying current. For large power transformers, the inrush current can persist for several seconds before it finally dies off.

The line reactance has the effect of reducing the peak inrush current by simply adding inductance to the air-core inductance of the winding. There is a definite relationship between inrush current and short-circuit current because both are related to the air-core inductance of the windings. (Remember that short circuits tend to exclude flux from the core.) Typically, a rule of thumb is that peak inrush currents are a little over 90% of peak short-circuit currents. The magnetic forces caused by inrush currents are generally much smaller than short-circuit forces, however. Because only one winding per phase is involved, there is no magnetic repulsion between windings.

The whole problem of analyzing exciting current inrush gets much more difficult when 3-phase transformers are involved. This is because the phase angles of the exciting voltages are 120° apart, there are interactions of currents and voltages between phases, and the three poles of the switching device do not close at exactly the same time. Nevertheless, it is safe to say that the peak magnitude of inrush current for three-phase transformers approaches the short-circuit current levels.

One of the interesting features of exciting current inrush is that since the current is fully offset, there are large percentages of *even* harmonics present. Even harmonics are otherwise rarely encountered in power circuits. There is also a phenomenon known as *sympathetic inrush*, where a transformer that is previously energized will exhibit a sudden change in current when a nearby transformer is switched on. Sympathetic inrush is caused by changes in line voltages from the inrush currents of the second transformer.

5.7 TANK OVERHEATING FROM ZERO-SEQUENCE CURRENTS

In Chapter 2, several differences between three-legged and five-legged core-form transformers were identified. In particular, there is no magnetic return path for zero-sequence flux in three-legged core-form transformers and if the core is excited by zero-sequence voltages, zero-sequence flux is produced that must return through the air, or through the air to the transformer tank. Since the reluctance of the air gap is large, the zero-sequence reactance is small, typically around 0.4 per unit of the transformer impedance base. If zero-sequence flux returns through the walls of transformer tank, there will be a net flux surrounded by the inner surface of the tank. This net flux induces a current induced in the tank wall in a direction opposite from the net ampere-turns in the transformer windings. This is illustrated in Figure 5.14.

In effect, the tank is a Δ-connected single-turn tertiary wound on each core leg. Since there is only one turn, the currents induced in it can be many times the zero-sequence currents in the actual transformer windings. Unfortunately the transformer tank is not designed to carry very much current and if the zero-sequence flux becomes too large, the induced currents in the tank can cause severe overheating in matter of seconds. In some cases, the temperatures get high enough to blister the paint on the tank.

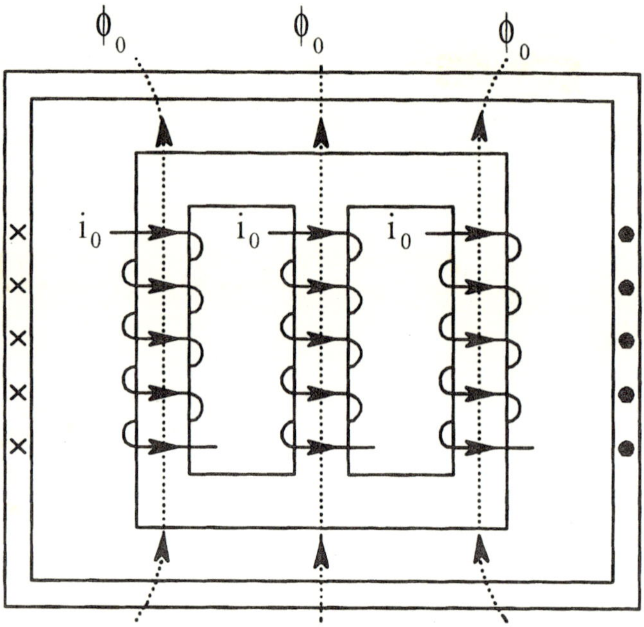

FIGURE 5.14 Tank currents in a 3-legged core form transformer excited with a zero-sequence current.

One common way for this condition to occur is with an open primary phase to a Grd.Y-Grd.Y transformer when a load connected to the secondary. This can easily happen in distribution systems where fuses are commonly used. The sequence network connections for an open *A* phase primary transformer connection is shown in Figure 5.15.

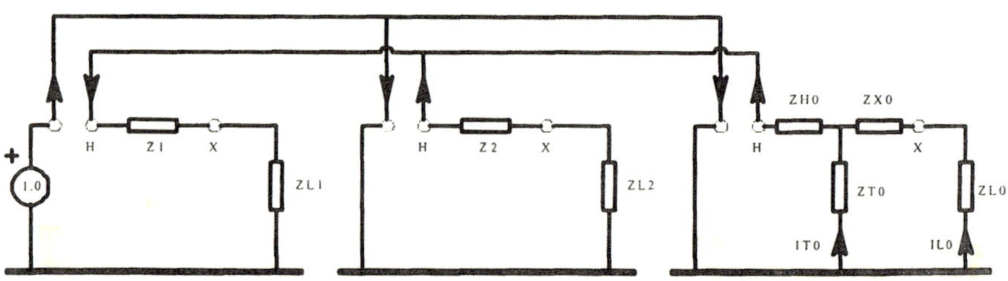

FIGURE 5.15 Positive, negative, and zero phase sequence networks connected for an open primary phase to a three-legged core form transformer carrying load.

In Figure 5.15, Z_{T0} is the magnetizing branch impedance of the zero-sequence equivalent circuit for the transformer. Z_{T0} is much smaller than the ordinary magnetizing impedance because of the high-reluctance path for zero-sequence flux in the three-legged core form design. I_{T0} is the zero-sequence magnetizing branch current. Note that I_{T0} would still exist even if there were no transformer tank present; however, the major portion of I_{T0} is represented by the current circulating in the tank.

Z_{T0} is typically around 30% of the transformer impedance base. The load impedances, Z_{L1}, Z_{L2} and Z_{L0}, should all be at least 100% of the transformer impedance base to prevent overloading the transformer. The series transformer impedances Z_1, Z_2, and $Z_{H0} + Z_{X0}$, are typically less than 10% of the transformer impedance base. With all of the above information and assuming that the load has a 90% power factor and the transformer impedances are purely reactive, I_{T0} is about 0.56 per unit for the circuit above. Since most of I_{T0} is tank circulating current, a typical value for tank currents under the conditions described above is about 50% of the transformer rating.

Example 5.4

A three-legged core form transformer is connected Grd.Y-Grd.Y with a 1 per unit load applied. One of the of the primary fuses blows open. Assume that the tank circulating current is 50% of the transformer rating, which is a fairly typical value. Calculate the tank circulating current in amperes for the following transformer characteristics.

KVA rating: 1500 KVA
Voltage rating: 12470Y/7200-208Y/120
Number of secondary turns: 10

Calculate the base current on the secondary side of the transformer:

$$I_{\text{base}} = \frac{(1{,}500{,}000 \text{ VA}/3)}{120 \text{ V}} = 4167 \text{ A per phase}$$

Since the transformer tank is equivalent to a Δ-connected tertiary surrounding *each* core leg by a single turn, the turns ratio between the secondary and the tank is 10:1. Calculating the tank current,

$$I_{\text{tank}} = 4167 \text{ A per phase} \times 10 = 41{,}670 \text{ A}$$

This example illustrates the magnitude of damaging currents that can occur for a Grd.Y-Grd.Y core-form transformer with a blown primary fuse. It is for this reason that this transformer design must be avoided when the possibility of an open primary phase exists.

5.8 PRIMARY FUSE MISOPERATIONS

We saw in Section 5.7 that tank overheating can easily result from a blown primary fuse. One of the drawbacks of protecting transformers with fuses is that they operate independently and somewhat randomly, which can result in some very serious consequences. Take, for example, a Δ-Grd.Y connected distribution substation transformer protected by primary fuses. For any phase-to-ground fault, there will be fault current through two of the primary fuses. If only one of the primary fuses blows, then the transformer will go into *single-phase operation* with two of the secondary phase voltages at normal voltage (but 180° apart in phase angle) and zero secondary voltage on the faulted phase. For single-phase loads, this is not a problem (except for loads connected to the faulted phase) but three-phase motors will not operate properly at these voltages and can easily be damaged. The situation can be analyzed using the method of symmetrical components; however, this requires applying two different faults simultaneously: an open-circuit fault at the primary plus a phase-to-ground fault at the secondary. It is more easily illustrated by the vector diagrams shown in Figure 5.16. If either the *A* phase primary fuse or the *B* phase primary fuse blows with a phase-to-ground fault sustained at the secondary, the voltage the *A* phase secondary winding is zero. This causes the voltage across primary winding between the *A* and *C* phases to collapse, resulting in the collapsed Δ shown in the vector diagrams on the right-hand side of Figure 5.16.

If the phase-to-ground fault clears, then the secondary voltage on one phase is normal and the secondary voltages of the other two phases will have one-half of the normal voltage, as illustrated by the vector diagram shown in Figure 5.17 for the case for a blown *A* phase primary fuse. The *A* phase voltage at the primary will be the average of the *B* phase voltage and the *C* phase voltage, collapsing the Δ as shown on the right-hand side of Figure 5.17. It is clear that only one-half of the normal voltages appear across the *A*-*B* primary winding and the *A*-*C* primary winding, and only one-half of the normal voltages appear across the *A* phase and *B* phase secondary windings. Any single-phase motors that are supplied by these phases will operate at half voltage, which will certainly cause motor damage if this conditions persists.

An even more serious situation can occur when a phase-to-phase secondary fault occurs on a Δ-Grd.Y connected transformer protected by primary fuses. There will be fault current in all three primary fuses; however, one of these fuses will have twice the fault current of the other two, and usually this

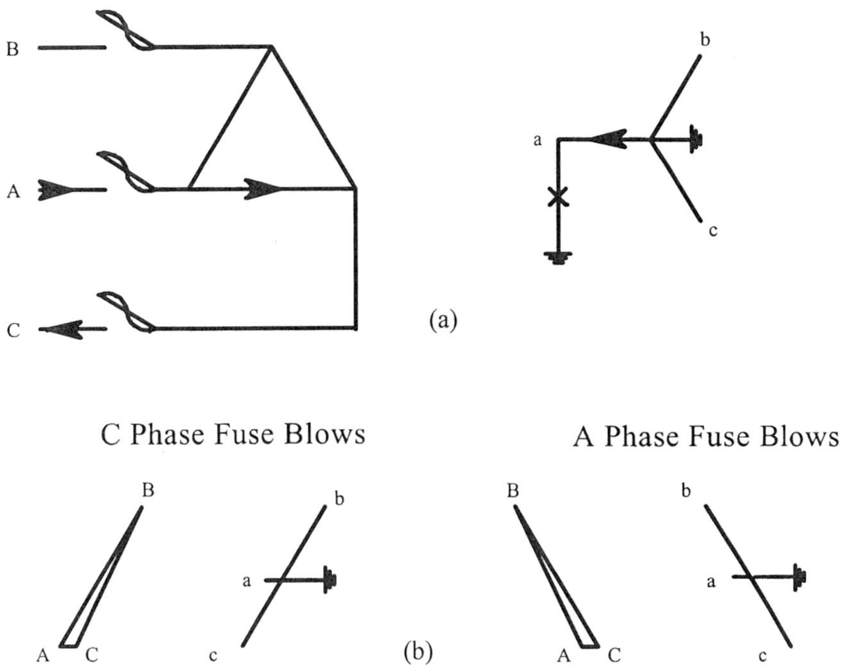

FIGURE 5.16 Vector diagrams of a phase-to-ground fault at the secondary of a Δ-Y transformer with one blown primary fuse. (a) Fault currents before fuse blows; (b) voltages after fuse blows.

fuse will blow first. This results in single-phase operation with normal secondary voltage on one phase and one-half normal secondary voltages on the other two phases.

On occasion, however, one of the other primary fuses will blow. If this happens, there will be normal secondary voltages on two of the phases and *double* the normal secondary voltage on one phase. Figure 5.18 shows this condition for a *B*-to-*C* phase secondary fault, which normally results in a blown *B* phase primary fuse. In this case, however, the *A* phase primary fuse blows instead.

Since *B* and *C* phase are shorted together on the secondary side of the transformer, the phase-to-neutral voltages are equal and so the *A*-*B* phase primary voltage must equal the *B*-*C* phase primary voltage. Since the *B* phase and *C* phase fuses are intact, the voltage across the *B*-*C* phase primary winding is the normal voltage. This forces the *A* phase voltage to be "outside the delta"

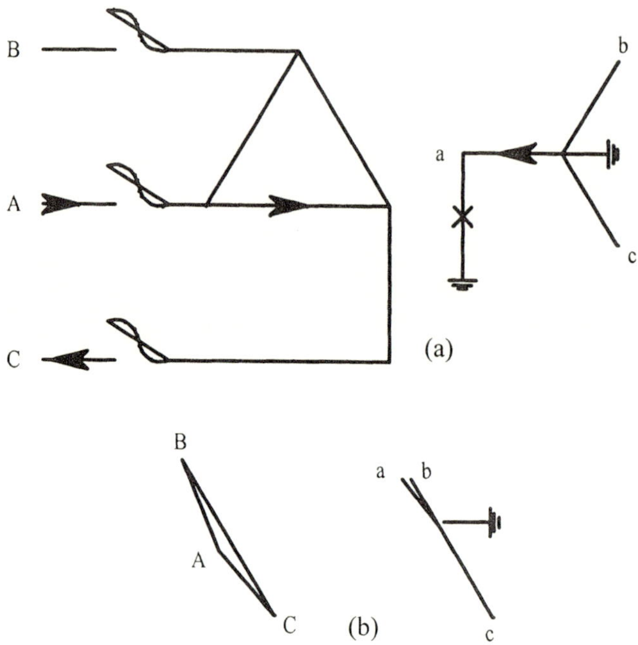

FIGURE 5.17 Vector diagrams of a phase-to-ground fault at the secondary of a Δ-Y transformer with one blown primary fuse after the fault clears. (a) Fault currents before fuse blows; (b) voltages after *A* phase blows, fault clears.

so the *A-B* voltage can be equal to, and in phase with, the *B-C* voltage. Since the *B-C* primary voltage is normal, the *A-C* primary voltage, which is equal to the sum of the *A-B* and *B-C* primary voltages, will be 2 times the normal voltage. The *A-C* primary voltage is induced across the *A* phase secondary winding as 2 times the normal secondary voltage. Consequently, the transformer operates in a single-phase mode with two of the secondary phases at normal voltage levels and the other secondary phase at 2 times the normal voltage level.

This is obviously a dangerous situation that can cause extensive equipment damage as well as bodily injury and possibly death to utility customers who are exposed to electric appliances operating at twice their normal voltages. In reality, a doubling of voltage is not quite achieved because the transformer core will saturate and thus limit the magnitude of the over voltage. Nevertheless, this type of fuse misoperation does occasionally occur, and the resulting equipment damage is usually quite extensive.

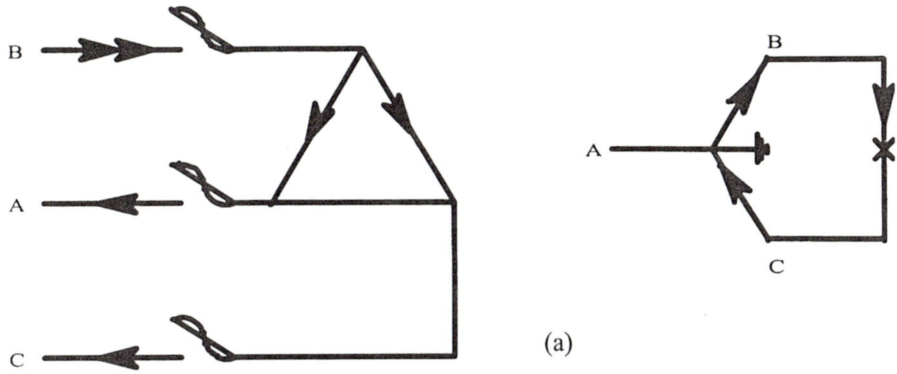

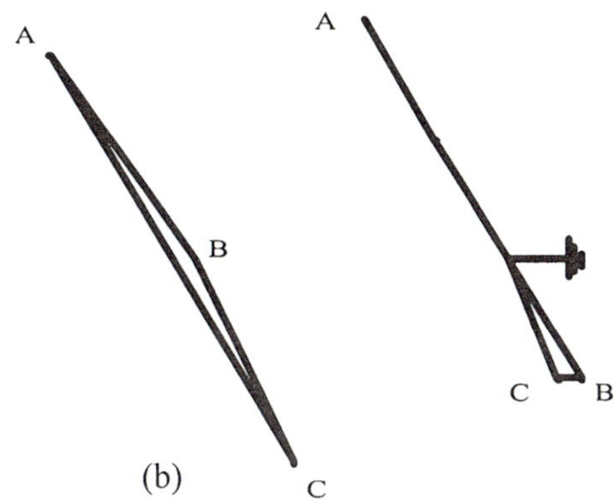

FIGURE 5.18 Vector diagrams of an *A* phase to *C* phase fault at the secondary of a Δ-Y transformer with the A phase fuse blown. (a) Fault currents before fuse blows; (b) voltages after *A* phase fuse blows.

5.9 FERRORESONANCE

Ferroresonance occurs when line capacitance resonates with the magnetizing reactance of a core while it goes in and out of saturation. Ferroresonance is usually associated with potential transformers, which are instrument transformers that are used to develop voltages used by relays; however, it can also occur with power transformers under special circumstances.

Ferroresonance is another occurrence that can cause equipment damage; fortunately, it is preventable by simply avoiding certain types of transformer connections with the types of circumstances that enable it to occur. Because these connections are routinely avoided in practice, ferroresonance is not encountered very often and there isn't much information about it in the literature. Ferroresonance is worthy of mention, however, because it can utterly destroy a transformer.

The necessary conditions for ferroresonance are established in the system shown in Figure 5.19. In the example shown in Figure 5.19, the Δ-connected tertiary winding of a large three-winding substation transformer

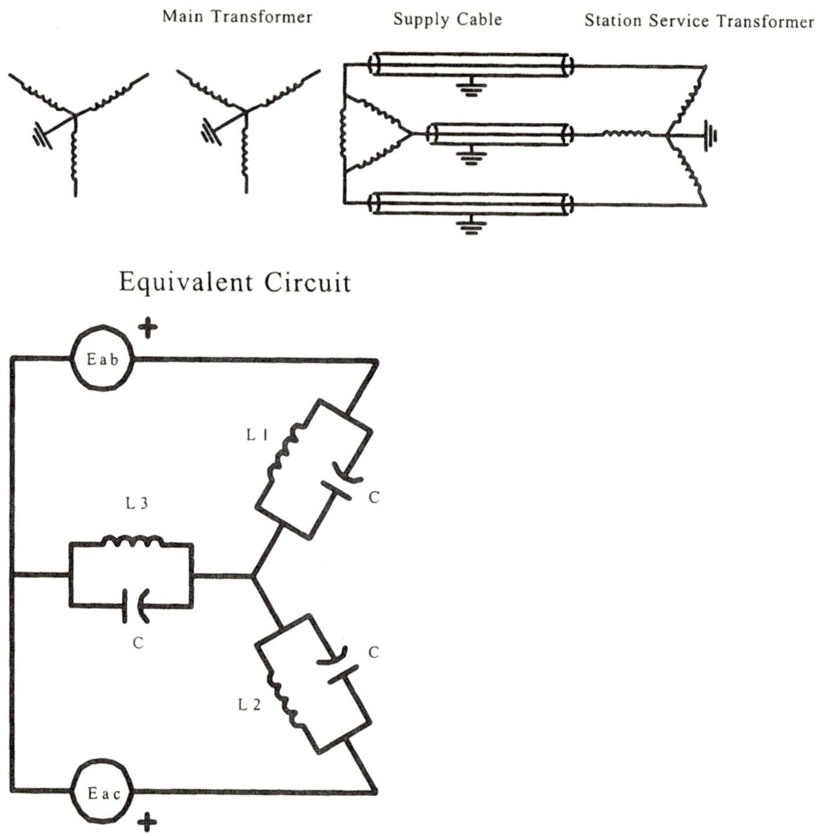

FIGURE 5.19 A situation where ferroresonance may occur. The cable capacitances form three parallel *L-C* elements that are in series with each other and the source voltages.

supplies a distribution type station-service transformer with a Grd.-Y primary winding. The supply lines to the station-service transformers are through a set of shielded cables. If the cable runs are fairly long, a significant amount of phase-to-ground capacitance may exist.

When a transformer core operates near saturation, the *B-H* curve is highly nonlinear, and the effective permeability of the core can take on a range of values that vary with the changes in flux density. Each of the inductances shown as L1, L2, and L3, will have instantaneous inductance values that are proportional to the effective permeability of the core at any given instant in time. These inductances form parallel *L-C* circuits that are in series with one another and in series with the source voltage. Since L1, L2, and L3 are constantly varying along with the effective permeability of the core, it is almost certain that a series resonant condition will exist at least part of the time during every cycle.

When a series resonance exists, even for a brief period, this causes very large voltages across the *L-C* elements. These voltages are capable of destroying the transformer and any other equipment connected to it. The nonlinear nature of this problem makes mathematical analysis virtually impossible, but the phenomenon has been observed both in the field and experimentally, and the voltages have been measured and recorded.

In the example above, the conditions for ferroresonance can be disrupted by the simple expedient of Δ-connected secondary winding to the station service transformer. The Δ-connected winding assures that the vector sum of the voltages of all three phases add to zero, stabilizing the neutral point of the Y-connected primary winding and preventing excessive voltage across the windings. The presence of a Δ-connected secondary winding will essentially "snuff out" ferroresonance in this circuit.

5.10 VOLTAGE SURGES

The effects of lightning and switching surges on a transformer are similar. The two properties that both of types of voltage surge have in common are: (1) a voltage peak that may be significantly greater than the normal line voltage, and (2) a *rate of change* of the voltage that is much faster than the rate of change of the normal line voltage. In fact, the rate of change alone can have more damaging effects on the turn-to-turn insulation of a winding than the magnitude of the peak voltage alone.

To study the effects of voltage surges, it is necessary to understand wave propagation through a transformer winding. A transformer winding is usually a coiled conductor with a series inductance between each turn and the turn

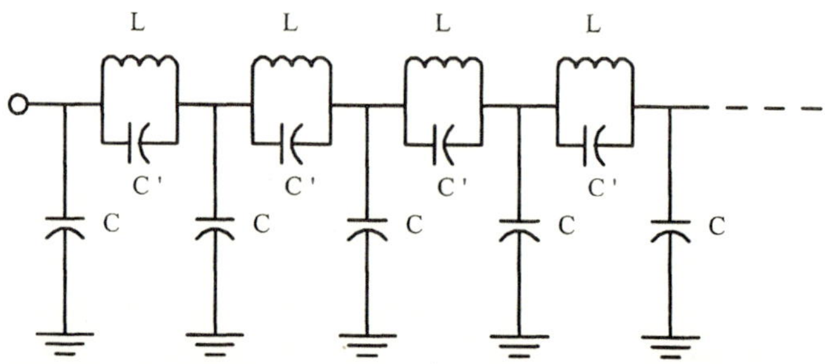

FIGURE 5.20 High-frequency model of a transformer winding. The inductance and capacitance elements form a waveguide that propagates voltage and current waves.

above and below it. Since the turns are coiled in close proximity with each other, there is also a shunt capacitance between each turn and the one above and below it. This forms a series of parallel *L-C* circuits from one end of the winding to the other, with each parallel *L-C* circuit representing one turn.

Since each turn is also capacitively coupled to ground, the winding takes on a structure similar to a waveguide, as illustrated in Figure 5.20, with a large number of parallel *L-C* elements in series and a large number of shunt capacitances to ground. Let

L = inductance per turn, H
C' = turn-to-turn capacitance per turn, F
C = capacitance to ground per turn, F

The high-frequency characteristics of the waveguide structure can be analyzed using Laplace transforms. For high-frequency phenomena, it is customary to ignore the transformer core, since the winding currents are highly localized. The core only comes into play at low frequencies where the winding currents are global.

In Laplace notation, the impedance of each L-C' element is represented by $Z(s)$, where s is the Laplace operator corresponding to the $\partial/\partial t$ operation. The admittance of each of the shunt elements C is represented by $Y(s)$.

$$Z(s) = \frac{sL}{1 - s^2LC'} \qquad (5.10.1)$$

$$Y(s) = sC \qquad (5.10.2)$$

When many of these elements are connected together, the resulting structure has a *characteristic admittance* $Y_0(s)$ and a *propagation constant*

$\gamma(s)$. In this case, unlike an ideal waveguide, the propagation "constant" is not really constant, but varies with frequency. Above a certain frequency ω_c called the *cutoff frequency*, the propagation constant becomes an *attenuation factor* and frequency components above ω_c will not propagate and decay exponentially instead.

$$Y_0(s) = \left[\frac{Y(s)}{Z(s)}\right]^{1/2} \quad \text{S} \tag{5.10.3}$$

$$\gamma(s) = [Y(s)Z(s)]^{1/2} \tag{5.10.4}$$

$$\omega_c = (LC')^{-1/2} \quad \text{rad/s} \tag{5.10.5}$$

Suppose a voltage surge is applied across the entire winding. This case occurs when a lightning surge arrives at the high-voltage terminals of a winding connected Grd.-Y. The voltage surge at the winding terminal is a time function $e_0(t)$ having a Laplace transform $e_0(s)$. The current at the winding terminal $i_0(t)$ also has a Laplace transform $i_0(s)$.

$$i_0(s) = e_0(s)Y_0(s) \tag{5.10.6}$$

Note that the current coming out of the winding at the grounded end is *not* the same as the current going into the winding at the line end because the voltage surge is a wave that must travel down the entire length of the winding before emerging at the grounded end. As the current wave passes through the winding, it is delayed by the propagation constant, so the current $i_x(s)$ at turn number x is related to the current at the line terminal by the following expression:

$$i_x(s) = e_0(s)Y_0(s)e^{-\gamma(s)x} \tag{5.10.7}$$

The *turn-to-turn voltage* $e_x(s)$ at turn number x is equal to the current $i_x(s)$ at turn number x times the impedance of the L-C' element $Z(s)$:

$$e_x(s) = e_0(s)Y_0(s)Z(s)e^{-\gamma(s)x} \tag{5.10.8}$$

but

$$Y_0(s)Z(s) = \left[\frac{Y(s)}{Z(s)}\right]^{1/2} Z(s) = [Y(s)Z(s)]^{1/2} = \gamma(s) \tag{5.10.9}$$

Therefore Eq. (5.10.8) reduces to the following expression:

$$e_x(s) = e_0(s)\gamma(s)e^{-\gamma(s)x} \tag{5.10.10}$$

The turn-to-turn voltage expressed as a time function $e_x(t)$ is solved by the inverse Laplace transform:

$$e_x(t) = \frac{-j}{2\pi} \int_{\sigma-j\infty}^{\sigma+j\infty} e_x(s) e^{st}\, ds \tag{5.10.11}$$

Equation (5.10.11) is not very difficult to solve numerically by computer. It is only necessary to numerically integrate the expression over a frequency range slightly greater than the cutoff frequency ω_c. The numerical integration is successfully carried out using a summation technique over the frequency range $-2\omega_c < \omega < 2\omega_c$ using the following parameters: $ds = j0.01\omega_c$ and $s = \sigma + j\omega$. The value of σ should be slightly positive. A value of $\sigma = 0.1\omega_c$ usually produces good results. This summation technique requires about 400 iterations for each value of $e_x(t)$ solved. Using complex variables this is a very easy task for a computer to perform.

The remaining task is to define the voltage surge and calculate its Laplace transform. A typical lightning surge is modeled as the so-called 1.2/50 μs full wave. This wave goes from zero to peak voltage in 1.2 μs and decays to 50% of peak value in 50 μs. The 1.2/50 μs full wave has the following time function:

$$e_0(t) = 8.333 \times 10^5 tu(t) - 8.4358 \times 10^5 tu(t - 1.2 \times 10^{-6}) \tag{5.10.12}$$

where $u(t)$ is the unit step function at $t = 0$.

The Laplace transform of $e_0(t)$ is

$$e_0(s) = \frac{(8.333 - 8.4358e^{-s\times 1.2\times 10^{-6}}) \times 10^{-5}}{s^2} \tag{5.10.13}$$

Plugging these functions into a computer algorithm that solves the inverse Laplace transform through a summing technique produces a plot of the turn-to-turn voltage as a function of time and as a function of the turn number x. The computer-generated plot shown in Figure 5.21 is for a 1.2/50 μs lightning wave propagating along a winding with $L = 1.0 \times 10^{-5}$ H per turn, $C = 1.0 \times 10^{-9}$ F per turn, and $C' = 1.0 \times 10^{-9}$ F-turn. The turn-to-turn voltages are shown at $x = 1$ and at $x = 20$ turns into the winding. Note that the turn-to-turn voltage is over 10% of the peak voltage that is applied *across the entire winding*.

A *chopped wave* also has a 1.2 μs wavefront, but the voltage then suddenly drops to zero as soon as the surge reaches peak voltage. The chopped wave simulates the effects of lightning arrestors that spark over and cause a

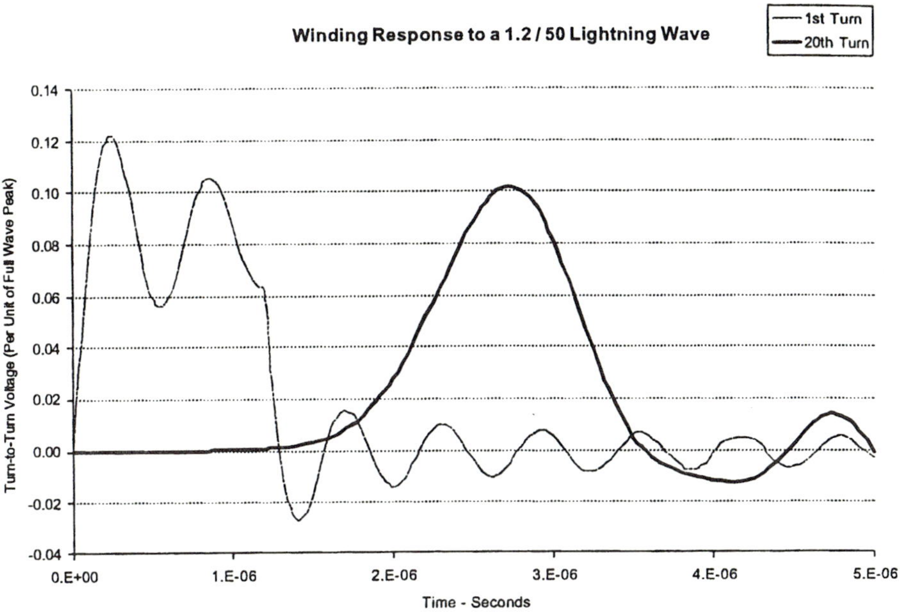

FIGURE 5.21 Computer simulation of the turn-to-turn voltages at the first turn and the twentieth turn in response to a full wave applied across the winding terminals. Note that the higher-frequency components above the cutoff frequency have been attenuated at the twentieth turn.

voltage collapse. The chopped wave can cause much more severe turn-to-turn voltages than the full wave because of the rapid change in voltage. The effects of the sudden voltage change are limited to some extent by the cutoff frequency ω_c which prevents the voltage from instantaneously changing. A computer simulation of the chopped wave voltages for the same winding used in the full-wave simulation (not shown here) results in turn-to-turn voltages approaching 50% of the peak voltage applied across the entire winding.

These extreme theoretical turn-to-turn voltages dissipate fairly rapidly as the wave moves along the winding; hence, the end turns of the winding usually show far more stress than the middle turns. If a transformer is designed to withstand a 750 kV chopped wave, the end turns would have to be able to withstand a turn-to-turn voltage of 375 kV!

Lightning-induced failures are typically manifested by turn-to-turn failures at or near the end turns of the windings. For a grounded winding, we would expect lightning-induced failures to occur most often occur at the line end, and they usually do; but some lightning-induced failures occur at the grounded end. This is explained by the fact that a traveling wave is reflected at the grounded end of the winding. The reflected line-to-ground voltage wave cancels the forward line-to-ground voltage wave, but the forward and reflected *current* waves add together and *double* the current at the grounded end. Since the *turn-to-turn voltage* is equal to the current times the impedance of the L-C' elements, the turn-to-turn voltage also doubles at the point of reflection.

Surge arrestors, also called lightning arrestors, are connected between the transformer terminals and ground to limit the peak surge voltage. Arrestors are coordinated with nominal system voltages so they will not spark over at normal line voltages. Basic impulse insulation level (BIL) ratings are based on the surge voltages that are expected to occur when arrestors are properly applied. Table 5.3 shows standard BIL ratings of power transformers for vari-

TABLE 5.3 Full Wave and Chopped Wave Test Voltages by BIL Class

BIL (kV)	Full wave test (kV)	Chopped wave test (kV)	Nominal system voltage (kV)
110	110	120	≤15
150	150	165	25
200	200	220	34.5
250	250	275	46–69
350	350	385	69–115
450	450	495	115–138
550	550	605	115–161
650	650	715	138–230
750	750	825	161–230
825	825	905	230
900	900	990	230–345
1050	1050	1155	345
1130	1130	1430	500
1175	1175	1290	345
1425	1425	1570	500
1550	1550	1705	500
1675	1675	1845	500
1800	1800	1980	765
1925	1925	2120	765
2050	2050	2255	765

Source: Ref. 1, p. 16.

ous nominal system voltages, and the voltage levels applied for the full-wave test and the chopped-wave test [1].

Example 5.5

A 150 MVA transformer has a 16% nameplate impedance with an X/R ratio of 10:1. The transformer is connected to a system impedance of 10% on a 100 MVA base with an X/R ratio of 5:1. What is the peak asymmetrical fault current in per unit for a three-phase fault on the load side of the transformer?

Converting the system impedance into the transformer base:

$$Z = \frac{0.10 \times 150 \text{ MVA}}{100 \text{ MVA}} = 0.15 \text{ per unit}$$

$$R^2 + X^2 = 0.15^2 = 0.0225 \text{ per unit}$$

Rearranging,

$$1 + \left(\frac{X}{R}\right)^2 = \frac{0.0225}{R^2}$$

Solving for the left-hand side of the equation,

$$1 + \left(\frac{X}{R}\right)^2 = 1 + 5^2 = 26$$

Substituting for $1 + (X/R)^2$:

$$26 = \frac{0.0225}{R^2}$$

Solving for R,

$$R^2 = \frac{0.0225}{26} = 0.0008654 \text{ per unit}$$

$$R = 0.0294 \text{ per unit}$$

Solving for X,

$$X = (R^2 + X^2) - R^2$$

$$= (0.0225 - 0.0008654)^{1/2} = 0.147 \text{ per unit}$$

For the transformer:

$$(R^2 + X^2) = 0.16^2 = 0.0256 \text{ per unit}$$

Rearranging,

$$1 + \left(\frac{X}{R}\right)^2 = \frac{0.0256}{R^2} = 1 + 10^2 = 101$$

Solving for the left-hand side of the equation,

$$1 + \left(\frac{X}{R}\right)^2 = 1 + 10^2 = 101$$

Substituting for the left-hand side of the equation,

$$101 = \frac{0.0256}{R^2}$$

Solving for R,

$$R^2 = \frac{0.0256}{101} = 0.0002535 \text{ per unit}$$

$$R = 0.0159 \text{ per unit}$$

Solving for X,

$$X = (R^2 + X^2) - R^2$$

$$X = (0.0256 - 0.0002535)^{1/2} = 0.159 \text{ per unit}$$

Combining system and transformer impedances,

$$\text{Total } R = 0.0294 + 0.0159 = 0.0453 \text{ per unit}$$

$$\text{Total } X = 0.147 + 0.159 = 0.306 \text{ per unit}$$

$$\text{Total } Z = (0.0453^2 + 0.306^2)^{1/2} = 0.309 \text{ per unit}$$

Calculating the total X/R ratio,

$$\frac{X}{R} = \frac{0.306}{0.0453} = 6.755$$

$$\phi = \tan^{-1}\frac{X}{R} = 1.424 \text{ rad}$$

Applying Eq. (5.2.2),

$$I_{\text{peak}}(\text{asym}) = 1.414 \times \frac{1 + e^{-(1.424+\pi/2)/6.755} \times \sin 1.424}{0.309}$$

$$= 7.482 \text{ per unit on a 150 MVA base}$$

Example 5.6

A 50 MVA single-phase shell-form transformer has a voltage rating of 79,670 to 39,840 V with an 8% nameplate impedance. The secondary winding is split into two pancake coils in series, each having 400 turns with the primary pancake coil sandwiched between the two secondary coils. The centers of the secondary windings are 10 cm from the center of the primary winding. A short circuit is applied to the secondary winding with the primary connected to a 1 per unit voltage source with zero impedance. Estimate the axial force on each of the coils. Assume that the magnetic centers of the coils are aligned perfectly in the radial direction and the magnetic centers each have an effective length of 3 m (see Figure 5.22).

Calculate the secondary short-circuit current:

$$I_f = \frac{50{,}000{,}000\ \text{VA}/39{,}840\ \text{V}}{0.08} = 15{,}688\ \text{A}$$

Secondary ampere-turns = 15,688 A × 400 turns = 6,275,100 ampere-turns, which are split into 2 halves. The primary ampere-turns = 6,275,100 ampere-turns.

The magnetic centers of the two secondary coils and the primary coil are 3 m long and are 0.1 m apart. All of the ampere-turns are concentrated at the magnetic centers as single turns, as shown in Figure 5.23. The forces on the primary coil cancel. The force on each secondary coil are calculated in MKS units:

$$F = \frac{6{,}275{,}100\ 1 \times 3{,}137{,}550/0.1 - 3{,}137{,}550 \times 3{,}137{,}550/0.2) \times 3 \times \mu_0}{2\pi}\ \text{N}$$

$$= 5.906 \times 10^6\ \text{N}\ (=1.328 \times 10^6\ \text{lbf in English units})$$

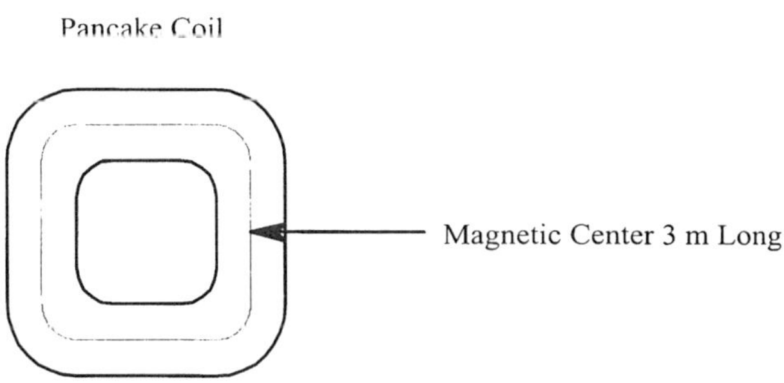

FIGURE 5.22 Pancake winding used in Example 5.6.

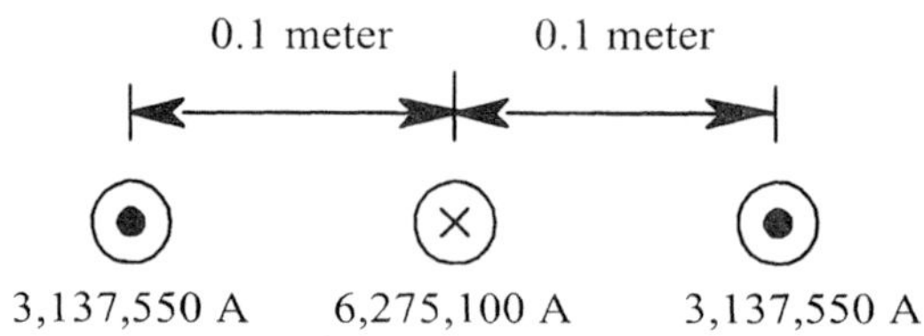

FIGURE 5.23 Location of the magnetic centers for Example 5.6.

Example 5.7

A Δ-connected tertiary winding of a large substation transformer supplies a Grd.Y-Δ station service transformer through a set of cables, as shown in Figure 5.24. The *A* phase cable sustains a phase-to-ground fault.

There are two possible fusing arrangements:

1. A set of fuses at the tertiary winding of the large substation transformer
2. A set of fuses at the primary of the station service transformer

Determine the phase-to-ground voltages at the station service transformer for both of the above fusing arrangements if the *A* phase fuse blows and the cable fault is sustained. Which of the fusing arrangements is preferable?

Arrangement A:

$$E_{AN} = 0 \qquad E_{CN} = \frac{12{,}470}{2} = 6235 \text{ V} \qquad E_{BN} = \frac{12{,}470}{2} = 6235 \text{ V}$$

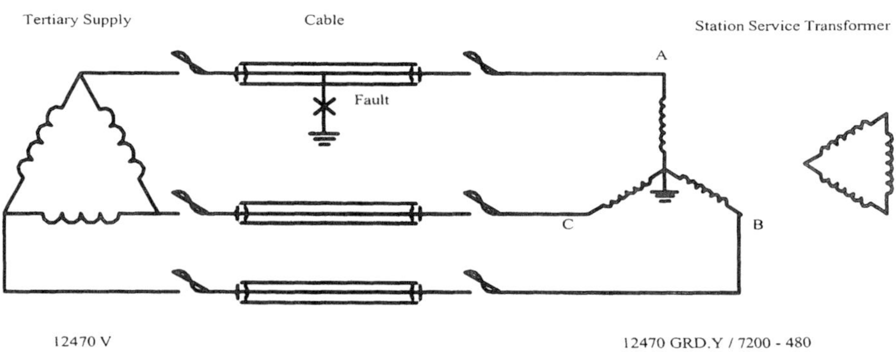

FIGURE 5.24 Circuit diagram for Example 5.7.

Arrangement B:

$$E_{CN} = E_{CA} = 12{,}470\ \text{V}\angle 210^\circ \quad E_{BN} = E_{BA} = 12{,}470\text{V}\angle 150^\circ$$

Due to the Δ secondary of the station service transformer, the sum of the voltages around the Δ must equal zero:

$$E_{AN} + E_{BN} + E_{CN} = 0$$

$$E_{AN} = -12{,}470\ \text{V}\angle 150^\circ - 12{,}470\ \text{V}\angle 210^\circ = 21{,}599\ \text{V}\angle 0^\circ$$

The rated phase-to-neutral voltage is 7200 V and the transformer is severely overvoltaged with arrangement B. Therefore, arrangement A is preferred.

REFERENCES

1. IEEE Std. C57.12.00-2000. IEEE Standard General Requirements for Liquid-Immersed Distribution, Power, and Regulating Transformers. New York: Institute of Electrical and Electronics Engineers, Inc., New York 2000.
2. R. L. Bean, N. Chakan, Jr., H. R. Moore and E. C. Wentz. Transformers for the Electric Power Industry. McGraw-Hill Book, New York, 1959.

6

Ancillary Equipment

Transformers are equipped with an array of ancillary equipment in order to survive and function in the field. This chapter describes how this equipment is designed, how it functions, and how it should be maintained. The proper design and maintenance of ancillary equipment is just as important as the proper design and maintenance of the transformer itself and can present its own set of problems, since unlike the transformer proper, ancillary equipment may contain moving parts that are subject to wear. Many transformer failures have been the result of carelessness with regard to ancillary equipment.

6.1 BUSHINGS

A transformer designed for outdoor use has its core and coils completely enclosed by a steel enclosure. In order to connect the windings to the electrical system, the leads are brought out of the tank through *bushings*. Since the leads are energized at line voltages, the bushings must insulate and isolate the leads from each other and from the tank. A cutaway view of a high-voltage oil-filled condenser bushing is shown in Figure 6.1.

A condenser bushing gets its name from the grading capacitor (aka condenser) that relieves the electrical stresses internal to the bushing. Core insula-

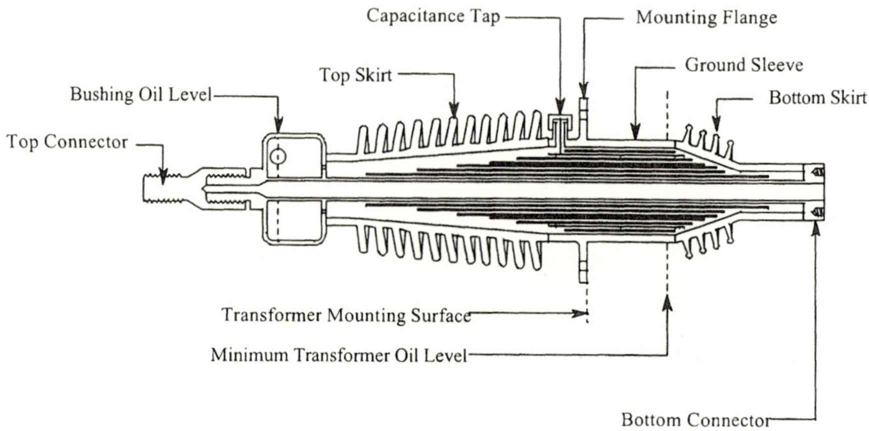

FIGURE 6.1 Cutaway sketch of a condenser bushing, shown on its side. (This bushing is normally mounted in the vertical position.)

tion, consisting of oil-impregnated kraft paper, and the conducting plates of the grading capacitor surround a hollow tube operating at line potential that runs through the middle of the bushing. A mounting flange that is at ground potential is bolted to the transformer tank with a ground sleeve extending below the flange. A skirted porcelain insulator is placed between the grounded mounting flange and the energized upper portion of the bushing. A shorter porcelain insulator is placed between the ground sleeve and the energized bottom portion of the bushing. An upper reservoir, with a sight glass or float oil level gauge, supplies insulating oil for impregnating the kraft paper insulation below. A capacitance tap is electrically connected to one of the plates of the grading capacitor for measuring the power factor of the insulation when the bushing is tested. This tap is normally grounded when the bushing is in operation.

In the configuration shown in the diagram, the hollow tube conducts current through the bushing with the transformer winding lead attached bolted to the bottom of the bushing through a connector. In a different configuration, called a *draw lead bushing*, the lead from the transformer winding is drawn through the hollow tube and terminates at the top of the bushing using a special type of top connector with inside threads that mate with the draw lead connector. The draw lead configuration generally has a lower ampere rating than the bottom-connector configuration because the draw lead conductor is usually smaller in cross-sectional area than the hollow tube. Also, there is an additional temperature gradient since I^2R losses from the draw lead must be transferred through the space between it and the inner surface of the hollow tube.

Skirts are used on the outer surfaces of the upper and lower insulators to lengthen the creepage paths between the energized ends of the bushing and the ground sleeve. Since porcelain is naturally a porous material, the outer surfaces of the upper and lower insulators are coated with a hard glaze to shed moisture and contamination. The oil level inside the transformer must be maintained above the bottom of the ground sleeve to assure that the lower porcelain insulator is immersed in oil. This allows the creepage path along the surface of the lower insulator in oil to be significantly shorter than the creepage path for the upper insulator in air.

EHV (extra high voltage) bushings have corona rings attached to the top of the bushing to prevent corona from the high voltage stresses at sharp edges and corners. Corona rings are generally used at voltages above 230 kV.

Porcelain is strong in compression and structurally weak in tension. Therefore, in order for a long bushing to withstand cantilever forces, the porcelain is compressed by means of massive springs in the top assembly that keeps the hollow tube in tension. Also, since it is impractical to fabricate the porcelain of a large bushing as a single assembly, large pieces are usually assembled as smaller sections that are interlocked and kept in compression.

Condenser bushings are also used for low-voltage bushing applications; however, *bulk-type* bushings are generally used for 15 kV and below. Bulk-type bushings consist of a porcelain insulator, a center conductor, and a mounting flange. There is no oil reservoir, and no grading capacitor. In some bulk-type bushing designs, the core insulation is made of kraft paper that is in communication with the transformer oil with wicking action to keep the paper oil-impregnated. In other bulk type bushing designs, all of the insulating material is made out of porcelain, which greatly simplifies the design. An example of a low-voltage bulk-type bushing design is shown in Figure 6.2. A solid conductor extends from the top and bottom connector through the center of the bushing. The insulation consists of a hollow porcelain tube with skirts on the upper portion only. The mounding flange and ground plane surrounds the middle of the porcelain insulator. Because the bushing operates at a low voltage, there is sufficient creepage path below the flange so the bushing can be operated safely without regard to a minimum oil level.

Bushings have voltage ratings, BIL ratings, and current ratings. Since bushings have no inherent overload capability, transformer loading is kept within the bushing ampere ratings and the bushings are generally sized somewhat larger than the nominal load in order to accommodate the transformer overload capability.

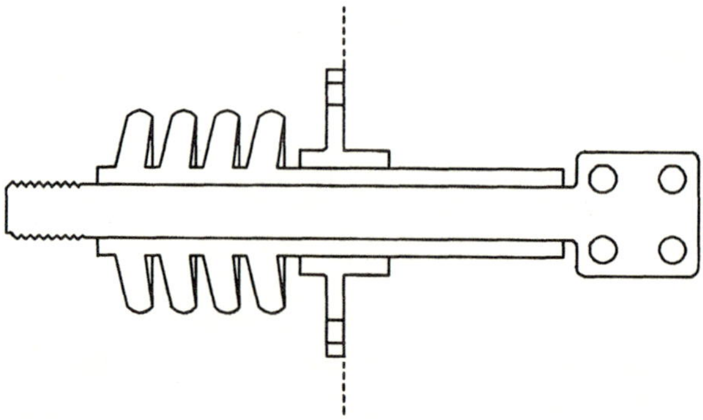

FIGURE 6.2 Cutaway sketch of a solid bushing, suitable for side-wall mounting. The internal connections are made to the four-hole pad on the right and the external leads are made to the threaded stud on the left.

Transformer bushings come in a bewildering array of shapes and sizes. For a long time, there was no effort on the part of transformer manufacturers to standardize bushings so that they could be interchanged among transformers of the same voltage classes and similar KVA ratings. Even when similar transformers were built by the same manufacturer, the bushings often would not be compatible with each other. To add yet more confusion, bushings supplied by the same manufacturer that were identical often carried different catalog numbers. The result was that each transformer had a set of bushings that was unique to that particular unit, and the transformer owners had to stock a large number of spare bushings to cover those in service.

Some utility companies instituted computer databases, entering the critical dimensions and other parameters for every in-service and spare bushing, so that matches could be found among bushings that were "close enough" in size and capacity. In recent years, there has been an effort to standardize bushings and for manufacturers to use standardized bushings in their transformers. As older transformers are culled from the population, bushing interchangeability should become much less of an issue in the future.

6.2 OIL PRESERVATION SYSTEMS

Transformer oil provides insulation as well as cooling for the core and coil assembly. Oil contamination is a serious problem because it robs the oil of its dielectric properties and renders it useless as an insulating medium. There-

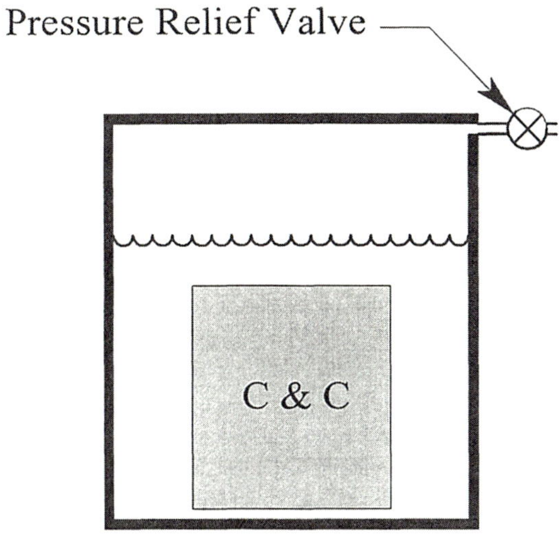

FIGURE 6.3 Sealed-tank oil preservation system.

fore, one of the most important design features of a transformer is the *oil preservation system*.

The simplest method of oil preservation is the *sealed-tank design*, as shown in Figure 6.3. The core and coils are immersed in oil with an air or gas blanket above the oil surface. The volume of the gas blanket must be sufficient to allow hot oil to expand inside the tank without over pressurizing the tank. Under extremes in temperature, the sealed-tank design may prove impractical because of the large gas volume that is required. A large volume of gas is generated during electrical faults. A pressure relief valve is provided to allow fault gases to escape In order to keep the tank from rupturing. The main advantages of the sealed-tank design is its inherent simplicity and low maintenance requirements. For these reasons, sealed-tank transformers are common with distribution transformers because they are seldom maintained.

One of the earliest alternative methods of oil preservation was the *conservator tank design*. In conservator-type transformers, the main tank is completely filled with oil with a smaller oil reservoir (conservator tank) mounted above the tank and connected to it by a pipe called a "gooseneck," as shown in Figure 6.4. The conservator tank is vented to the atmosphere at the top and the normal oil level is approximately in the middle of the conservator to allow for expansion and contraction. The purpose of the gooseneck is to break up the convection flow of hot oil from the transformer tank and the cooler oil in

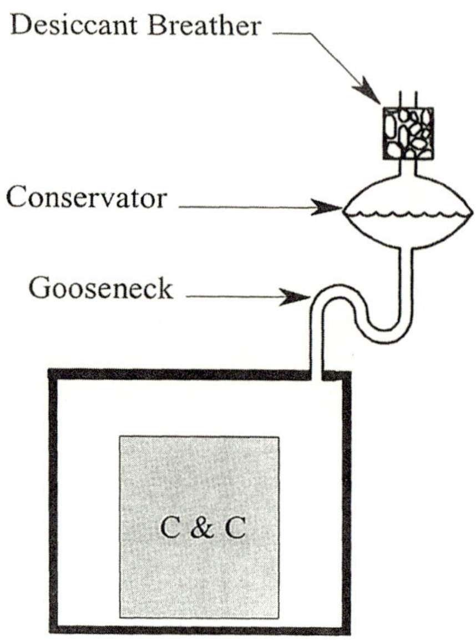

FIGURE 6.4 Conservator oil preservation system.

the conservator tank. This is only partially successful, however, so there is always some communication between the transformer oil bathing the core and coils and the oil in the conservator tank that is exposed to the atmosphere. The oil in the transformer tank is under a slight positive pressure equal to the head of oil above it. Although the conservator tank design prevents over pressurization when the oil expands, it is only marginally effective in keeping the oil free of contaminants from the atmosphere. One variation uses a desiccant canister in the breathing vent to absorb moisture, which keeps the atmosphere inside the conservator tank dry. The desiccant canister requires replenishment at intervals. A color change from pink to blue indicates when the desiccant needs to be replaced.

A variant of the conservator tank design places an expandable bladder between the oil and the outside air. This design has been fairly successful over the years, although the bladder can be damaged or torn.

Another oil preservation system that is a variation of the conservator is called the *gas oil seal* design, as depicted in Figure 6.5. A gas oil seal system has a relatively small gas blanket of inert nitrogen over the oil in the main transformer tank. Dissolved oxygen in the oil leads to a chemical reaction

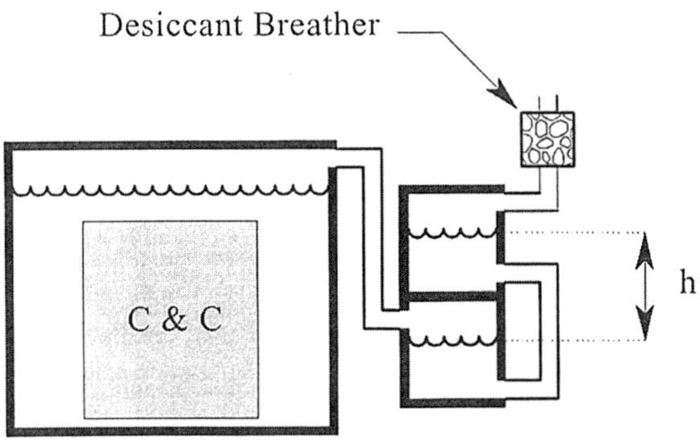

FIGURE 6.5 Gas-oil-seal oil preservation system.

known as oxidation. The byproducts of oxidation include H_2O, acids and sludge, all of which will shorten the life of the transformer, along with various hydrocarbon compounds. Using a blanket of inert nitrogen over oil that has been degassed will considerably reduce dissolved oxygen and oxidation by-products.

The gas space over the core and coils is connected to the gas space of a lower sealed oil reservoir. The oil in the bottom of the lower reservoir communicates with oil in the bottom of an upper reservoir. A blanket of air, which communicates with the atmosphere, covers the oil in the upper reservoir. A head of oil h exists between the oil levels in the upper and lower reservoirs. A positive pressure in the lower reservoir must maintain this head of oil, and hence there must be a positive pressure in the main transformer tank as well.

As the oil in the main tank expands thermally, the increased pressure of the nitrogen blanket forces oil out of the bottom of the lower reservoir into the bottom of the upper reservoir. This increases the head of oil h balancing the increased pressure. The oil in the lower reservoir ''seals'' the nitrogen blanket in the main transformer, hence the name, and keeps the oil in the main tank free from contamination. In some systems, a dehydrating breather, filled with desiccant, further minimizes moisture intrusion into the oil. The main disadvantage of this system is that if there are any gas leaks in the main tank, the positive pressure cannot be maintained and the oil in the upper reservoir drains into the lower reservoir, collapsing the head of oil.

One other major oil preservation system has been called by various brand names, e.g., Inertair supplied by Westinghouse Electric. This system, depicted in Figure 6.6, uses a pressurized nitrogen bottle with a pressure

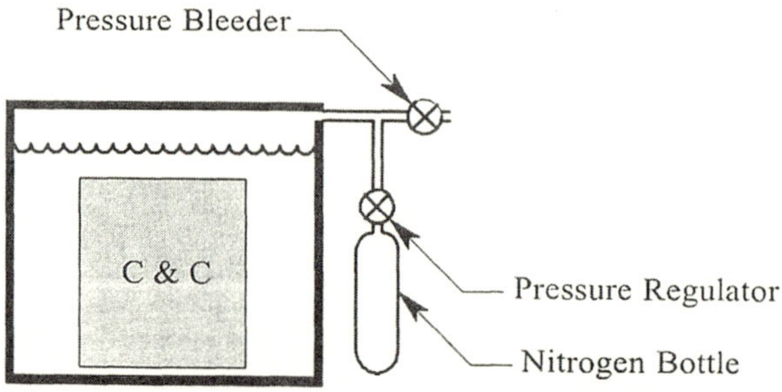

FIGURE 6.6 Inertaire oil preservation system.

regulator/breather that maintain a nearly constant positive pressure in a nitrogen blanket over the oil in the main tank. The main advantage of the regulated pressure design is that it is almost impervious to oil contamination and will perform its function even if there are small gas leaks. The disadvantage is that the nitrogen bottle requires periodic replacement, so maintaining the system is somewhat expensive. The nitrogen bottle will require frequent replacement if there is excessive temperature cycling or if there are significant gas leaks.

6.3 COOLING EQUIPMENT

Heat from core losses and copper losses must be dissipated to the environment. In dry type transformers, cooling is accomplished simply by circulating air around and through the coil and core assembly, either by natural convection or by forced air flow from fans. This cooling method is usually limited to low-voltage indoor transformers (5 kV and below) having a three-phase rating below 1500 KVA. At higher voltages, oil is required to insulate the windings, which prevents the use of air for cooling the core and coils directly. At higher KVA ratings, the losses are just too high for direct air cooling to be effective. In outdoor environments, direct air cooling would introduce unacceptable amounts of dirt and moisture into the windings.

Transformers come in various *cooling classes*, as defined by the industry standards. In recent years, there have been attempts to align the designations that apply to transformers manufactured in North America with the IEC cooling-class designations. Table 6.1 [1] gives the IEC designations and the earlier designations that are used in this book. All of the IEC designations use four letters. In some respects, the IEC designations are more descriptive than the North American designations because IEC makes a distinction between forced-oil/air cooled (OFAF) and directed-flow-air cooled (ODAF). Some

TABLE 6.1 Designations and Descriptions of the Cooling Classes Used in Power Transformers

Previous designation	IEC designation	Description
OA	ONAN	Oil-air cooled (self-cooled)
FA	ONAF	Forced-air cooled
OA/FA/FA	ONAN/ONAF/ONAF	Oil-air cooled (self-cooled), followed by two stages of forced-air cooling (fans)
OA/FA/FOA	ONAN/ONAF/OFAF	Oil-air cooled (self-cooled), followed by one stage of forced-air cooling (fans), followed by 1 stage of forced oil (oil pumps)
OA/FOA	ONAN/ODAF	Oil-air cooled (self-cooled), followed by one stage of directed oil flow pumps (with fans)
OA/FOA/FOA	ONAN/ODAF/ODAF	Oil-air cooled (self-cooled), followed by two stages of directed oil flow pumps (with fans)
FOA	OFAF	Forced oil/air cooled (with fans) rating only—no self-cooled rating
FOW	OFWF	Forced oil/water cooled rating only (oil/water heat exchanger with oil and water pumps)—no self-cooled rating
FOA	ODAF	Forced oil/air cooled rating only with directed oil flow pumps and fans—no self-cooled rating
FOW	ODWF	Forced oil/water cooled rating only (oil/water heat exchanger with directed oil flow pumps and water pumps)—no self-cooled rating

Source: Ref. 1.

people find using the four-letter designations somewhat awkward, and this book uses the earlier designations throughout.

In small oil-filled distribution transformers, the surface of the tank is sufficient for transferring heat from the oil to the air. Ribs are added to the tanks of some distribution transformers to increase the surface area of the tank and to improve heat transfer. Large distribution transformers and small power transformers generally require radiator banks to provide cooling. Regardless of whether the tank surface, ribs, or radiators are used, transformers that transfer heat from oil to air through natural convection are all cooling class OA transformers.

Radiators used on OA transformers generally have round cooling tubes or flat fins with large cross section areas in order to allow oil to flow by natural convection with minimal resistance. Hot oil from the core and coils rises to the top of the tank above the inlet to the radiator. Cool oil from the radiator sinks to the bottom of the radiator through the outlet and into the bottom of the core and coils. This process is called thermo-siphoning and the oil velocity is relatively slow throughout the transformer and radiators. For this reason, OA transformers have relatively large temperature gradients between the bottom oil and the top oil, and relatively large temperature gradients between the winding temperatures and the top oil temperature. Likewise, the air circulates through the radiator through natural convection, or is aided by the wind.

As the transformer losses increase, the number and size of the radiators that are required to cool the oil must increase. Eventually, a point is reached where wind and natural convection are not adequate to remove the heat and air must be forced through the radiators by motor-driven fans. Transformers that have forced air cooling are cooling class FA transformers. FA transformers require auxiliary power to run the fan motors, however, and one of the advantages of OA transformers is that they require no auxiliary power for cooling equipment. Since additional cooling is not usually needed until the transformer is heavily loaded, the fans on most FA transformers are turned off until temperatures exceed some threshold value, so under light load the transformer is cooled by natural convection only. These transformers are cooling class OA/FA transformers.

Some transformers are cooled by natural convection below temperature T_1, turn on one stage of fans at a higher temperature T_2 and turn on a second stage of fans at an even higher temperature T_3. These transformers are cooling class OA/FA/FA transformers. The direction of air flow in forced-air units is either horizontally outward or vertically upward. The vertical flow pattern has the advantage of being in the same direction as the natural air convection, so the two air flows will reinforce each other.

Although the cooling capacity is greatly increased by the use of forced air, increasing the loading to take advantage of the increased capacity will increase the temperature gradients within the transformer. A point is reached where the internal temperature gradients limit the ability to increase load any further. The solution is to increase the oil velocity by pumping oil as well as forcing air through the radiators. The usual pump placement is at the bottom of the radiators, forcing oil from the radiator outlets into the bottom of he transformer tank in the same direction as natural circulation but at a much higher velocity. Such transformers are cooling class FOA transformers. By directing the flow of oil within the transformer windings, greater cooling efficiency can be achieved. In recognition of this fact, the calculation of hot-spot temperatures is modified slightly for directed-flow cooling class transformers.

As in forced-air designs, forced-oil cooling can be combined with OA cooling (OA/FOA) or in two stages (OA/FOA/FOA). A transformer having a stage of fans and a stage of oil pumps that are switched on at different temperatures would be a cooling class OA/FA/FOA transformer.

The radiator design on FOA transformers can differ substantially with the radiator design on FA transformers. Since the oil is pumped under considerable pressure, the resistance to oil flow is of secondary importance so the radiator tubes can be designed to maximize surface area at the expense of cross section area. FOA radiators are sometimes called coolers instead, and tend to resemble automotive radiators with very narrow spaces between the cooling tubes and flat fins in the spaces between the cooling tubes to provide additional surface area. The comparison of the two types is illustrated in Figures 6.7 (OA/FA type) and 6.8 (FOA type).

FIGURE 6.7 OA/FA radiator construction. The large radiator tubes minimize restriction of oil flow under natural convection. The fan is shown mounted at the bottom with air flow directed upward.

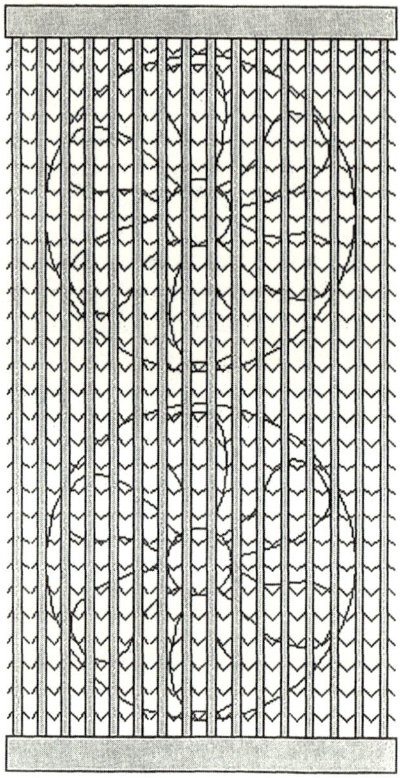

FIGURE 6.8 FOA cooler construction. The oil is forced through narrow tubes from top to bottom by means of oil pumps. The cooling fans direct air horizontally outward.

Cooling equipment requires maintenance in order to run efficiently and provide for a long transformer life. There is the obvious need to maintain the fans, pumps, and electrical supply equipment. The oil coolers themselves must be kept clean as well, especially FOA-type coolers. Many transformers have overheated under moderate loads because the cooling fins were clogged with insect and bird nests, dust, pollen, and other debris. For generator step-up transformers, where the load is nearly at nameplate rating continuously, steam-cleaning the coolers once every year is a good maintenance practice.

6.4 TEMPERATURE, OIL LEVEL, AND PRESSURE GAUGES

Temperature controls are required in order to turn on and off the cooling equipment. These controls are usually combined with a visible dial-type gauge that measures "winding temperature." This is really a misnomer because the gauge does not actually contact the winding. Instead, it measures the top oil temperature plus a temperature gradient produced by a small heater surrounding the thermometer bulb. This heater is connected to a current transformer on one of the phases of the secondary leads, so as the secondary load increases, so does the current through the heater resistance. The heater then becomes a mimic of the actual transformer winding with the winding temperature gauge measuring a temperature that is roughly equivalent to the true winding temperature.

In addition to (or instead of) the winding temperature gauge, most transformers have an oil temperature gauge that measures the actual top oil temperature. The difference in temperature between the winding temperature gauge and the top oil temperature gauge is one indication of how heavily the transformer is loaded. Winding and/or top oil temperatures should be routinely observed to see whether the transformer is operating within normal temperature limits. There have been many cases where an abnormal temperature indication has uncovered serious problems with transformers.

One example of this was a case of an OA cooling class transformer operating with a very high oil temperature. Further investigation revealed that a leak in the radiator had caused the oil level to drop below the radiator inlet, resulting in a virtual total loss of cooling. In other cases, investigations of elevated transformer temperatures have revealed blocked coolers and malfunctioning cooler controls.

An oil level gauge is required so that the correct oil level can be maintained. There is usually a mark on the gauge that indicates the 25°C level, which is the proper oil level at that temperature. Maintaining the proper oil level is extremely important because if the oil level falls below the level of the radiator inlet, flow through the radiator will cease and the transformer will overheat. A very low oil level can expose energized and current-carrying components that are designed to operate in oil and could result in overheating or an electrical flashover. If the oil level is too high, it could cause over pressurization when the oil expands.

If the transformer is equipped with a constant-pressure nitrogen oil preservation system, then pressure gauges are included to measure the pressure of the gas blanket and the nitrogen in the make-up bottle.

Transformers that have a gas blanket over the oil in the main tank usually come equipped with pressure gauges. On random inspections, a sealed-tank unit should have a slight positive or a slight negative pressure. If the pressure gauge consistently records zero pressure under all loading and temperature conditions, this is an indication that the transformer tank has a leak, allowing it to "breathe." This condition must be taken seriously and fixed, because a transformer that breathes is probably contaminated with moisture.

6.5 TAP-CHANGING EQUIPMENT

A transformer's rated voltage may not match the system voltage exactly, or it may be necessary to raise or lower the output voltage to supply a certain load. In these cases, a portion of a winding can be removed or added to change the transformer turns ratio. The simplest tap-changing device operates on a "break before make" principle and changes taps on the primary winding. Obviously, such a device cannot be operated when the transformer is carrying load or even when the transformer is energized because it would break load current and/or magnetizing current otherwise. This device is called a tap changer for deenergized operation. Years ago, it was called a no-load tap changer, but this description has fallen out of favor because the name implies that it can be operated when the transformer is energized but not carrying load, which is *not* the case. Most tap changers for deenergized operation have a total of five tap positions. There are usually two tap positions above the nominal voltage rating and two tap positions below the nominal voltage plus a tap at the nominal voltage. The voltage increments between taps are generally $2^1/_2\%$ of the nominal voltage, so the full tap range is $\pm 5\%$.

A rotary drum-type tap changer and the winding connections are shown in Figure 6.9. An insulated rotating shaft aligns a spring-loaded shorting wedge between pairs of hollow cylindrical electrodes. The winding taps are connected to the cylindrical electrodes by soldering or brazing. Different manufacturers have used other rotary tap changer configurations and mechanisms with linear movement are available. The tap changers for deenergized operation are usually mounted above the core and coil assembly to make inspection and maintenance possible. The taps are changed by a wheel on the outside of the transformer that locks in each tap position with indicating markers to show what tap the transformer is on. In three-phase transformers, the tap changers of all three phases are mechanically ganged together so that all three phases are always on the same taps.

If the tap changer is in the primary winding, then the primary current increases with respect to the load current on the lower taps, increasing the

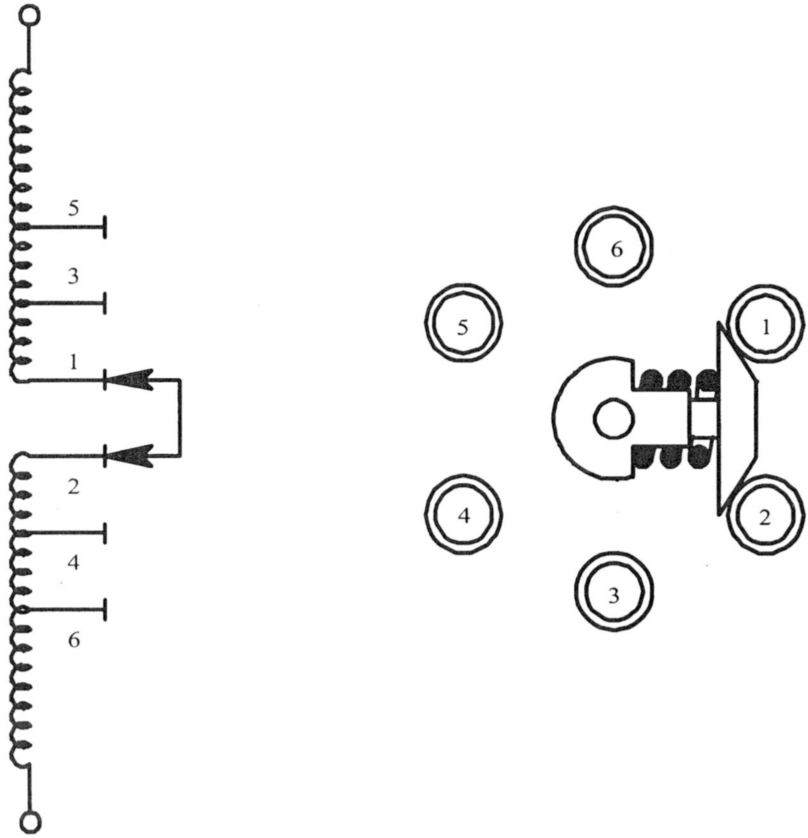

Figure 6.9 A tap changer for deenergized operation (break before make). There are five tap positions, with the connections for full winding taps shown

primary winding current with respect to the load current. The fact that part of the winding resistance is removed from the circuit on the lower taps partially compensates for the increased current; however, the lowest tap still has the highest total I^2R losses for a given load current. The winding taps are always in the middle portion of the winding to keep the magnetic center aligned with the other windings as sections of the winding are removed from the circuit by changing taps. Since the winding is never quite balanced on two of the five taps, the winding can be split into three sections, as shown in Figure 6.10. This arrangement requires two tap-changing mechanisms per phase, but the

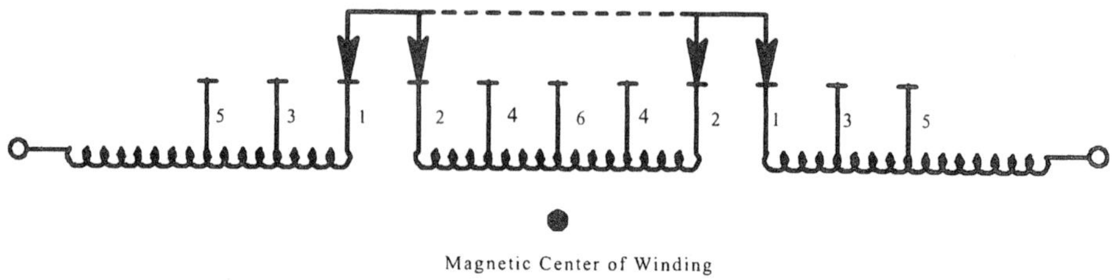

FIGURE 6.10 A tapped winding with two sets of taps that balance the ampere-turns around the geometric center of the winding. This configuration minimizes the vertical forces when the transformer is subjected to short circuits.

magnetic center of the winding is always aligned at the number 6 tap to help control short-circuit forces.

Tap changers for deenergized operation are designed to be moved infrequently. The tap settings are generally specified for the particular location on the electrical system and the settings do not change unless system conditions permanently change. Because a good electrical contact often depends on contact "wiping," it is generally a good idea to operate the tap changer periodically (when the transformer is out of service) to wipe the contacts clean. The contacts themselves are generally silver- or tin-coated since bare copper has a tendency to develop a copper sulfate film under oil which increases the contact resistance. This can lead to a thermal runaway effect from *oil coking*. As the temperature around the contacts increases, the oil around the contacts can coke or turn into carbon. This forms a carbon film that can actually force the contacts apart so the load current must pass through a layer of high-resistance carbon. This increases the temperature still further, leading to more coking and so forth until the contacts overheat and are destroyed.

After a tap changer is operated, it is important that the tap position is verified by a transformer turns ratio (TTR) test. The details of the TTR test will be covered in Chapter 8. There have been numerous cases where the tap changers of one or more phases were not correctly aligned even when the external markings indicated that they were. Energizing a transformer with tap changer contacts parted will result in immediate failure when the winding current arcs across the gap.

6.6 LOAD TAP CHANGERS

When load levels and/or system voltages change frequently, it is sometimes necessary to adjust the transformer tap ratio to follow the changes in system conditions. It is obviously impractical to do this while the transformer is deenergized or unloaded, so a special type of tap changer has been developed to change taps under full-load conditions.

Whereas the tap changer for deenergized operation is a break-before-make switching device, a load tap changer (LTC) must be a make-before-break switching device, requiring bridging over two adjacent taps before moving on to the next tap. If an electrical short circuit were placed between two taps, then the short-circuit current would be extremely large based on the large number of ampere turns with few turns. Therefore, an impedance must be inserted between the taps in order to limit the short-circuit current that flowing in the bridging position.

Transformer manufactures in North America adopted the *preventative autotransformer* method as the solution to this problem. This scheme is depicted Figure 6.11, showing the bridged position. Note that the tapped winding is *not* magnetically coupled with the preventative autotransformer. The preventative autotransformer has a 1 : 1 turns ratio and with the tap changer in the bridged position and the ampere-turns in each half of the preventative autotransformer must cancel each other. Therefore, the load current flows virtually unimpeded through both halves of the autotransformer winding. Any circulating current that flows in the closed loop between the taps encounters the magnetizing impedance of the preventative autotransformer, which effectively blocks the circulating current. The voltage at the output of the preventative autotransformer is the average of the voltages of the bridged taps.

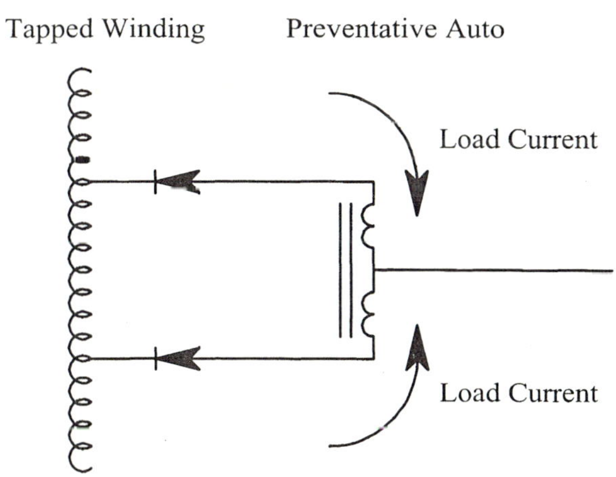

FIGURE 6.11 A load tap changer (LTC) in the bridging position with load currents evenly divided between the movable contacts.

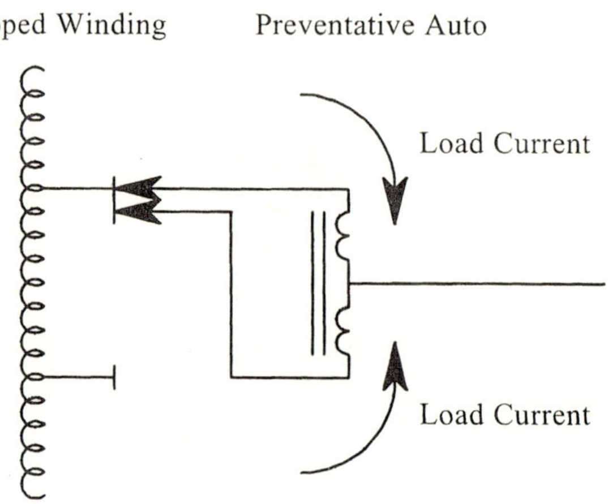

FIGURE 6.12 A load tap changer (LTC) in the full tap position.

The tap changer in the unbridged, full-tap position is illustrated in Figure 6.12. With the tap changer in this configuration, all of the load current is coming from a single tap through both moving contacts and the output voltage is equal to the voltage at that tap. Again, the ampere-turns in the two halves of the 1 : 1 preventative autotransformer cancel, so the load current flows virtually unimpeded.

When the tap changer begins to move to the next tap position, the upper moving contact must part while the lower moving contact remains on the tap. All of the load current in the upper moving contact is then suddenly transferred to the lower moving contact, as shown in Figure 6.13. Since all of the load current now flows through one-half of the autotransformer winding, it encounters the magnetizing impedance of the autotransformer, which would normally create a large voltage drop. This problem is remedied by designing the preventative autotransformer with a relatively low magnetizing impedance by building in air gaps in the core and by allowing the core to saturate at a relatively low winding voltage. The saturation voltage should be slightly higher than the voltage between taps in order to prevent circulating current in the bridged position yet low enough not to introduce an excessive voltage drop at the load terminals.

During the load transfers between moving contacts, there is considerable arcing due to the very large series inductance in the tapped winding and the windings of the preventative autotransformer. It is therefore necessary to perform a load transfer through a set of stationary arcing contacts in order to

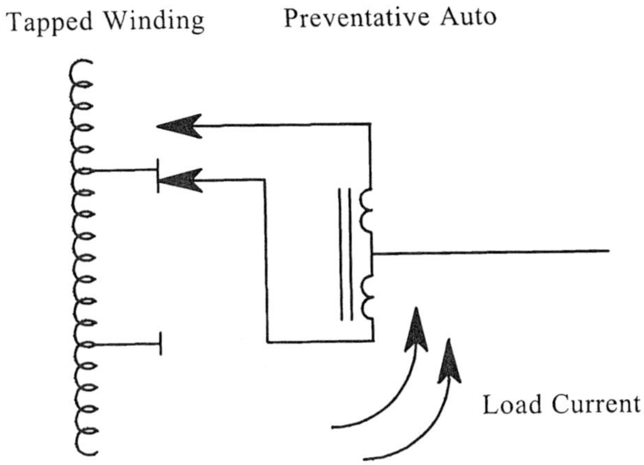

FIGURE 6.13 A load tap changer (LTC) beginning to move to the bridging position.

prevent burning and pitting of the moving contacts when breaking load current. One of the many schemes for performing the load transfer is illustrated in Figure 6.14.

When both moving contacts are on the same tap, all of the arcing contacts are closed. Before the upper moving contact moves, the upper transfer contact opens. Since the middle arcing contact and the lower transfer contact

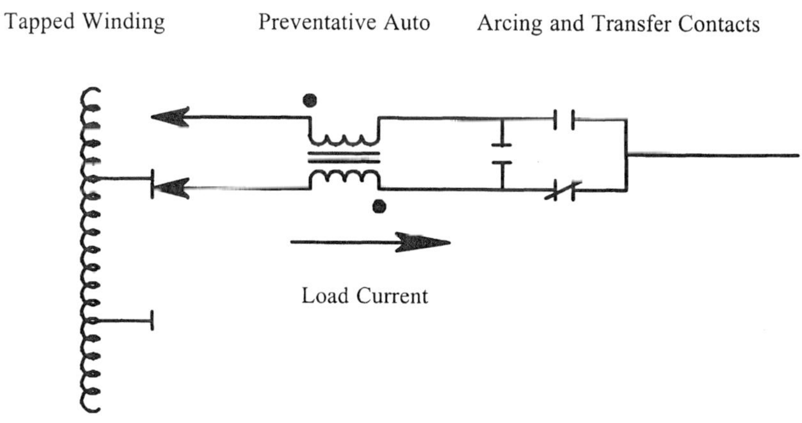

FIGURE 6.14 A load tap changer (LTC) beginning to move to the bridging position showing the load current through a single set of transfer contacts.

are closed, the upper transfer contact is merely breaking a parallel current, so minimal arcing occurs there. Next, the middle arcing contact opens, interrupting the load current through the upper moving contact in series with the transformer winding inductances. This produces a substantial amount of arcing and the middle arcing contacts must be designed to withstand this duty. Finally, with the upper moving contact isolated, it can move on to the next tap position. When the upper moving contact is established on the next tap, all of the arcing and transfer contacts are closed to establish the bridging position. The movement of the lower moving contact can proceed in a similar manner, first by opening the lower transfer contact, then the middle arcing contact, and finally the lower moving contact. Notice that this scheme requires only one set of arcing contacts for both sets of moving contacts.

The main advantage of the preventative autotransformer scheme is that the tap changer can stay on the bridging position indefinitely. Thus, bridging positions can be utilized the same as fixed taps, reducing the number of physical taps required to provide the required number of voltage changes. The main disadvantage of this method is that a separate set of core and coils must be provided for the preventative autotransformer. In a three-phase transformer with load tap changers, one three-phase preventative autotransformer is used instead of three separated preventative autotransformers for economy. Since the KVA transformed is relatively small, the preventative autotransformer can usually be accommodated within a fairly small space inside the main tank. LTCs are generally installed on the secondary side of the transformer.

LTCs require periodic maintenance. The motor-driven mechanism needs to be cleaned and lubricated and the controls inspected and tested. The main moving contacts should be periodically inspected and replaced if worn or pitted. The arcing contacts are the components that have the greatest duty and require the most frequent overhaul. If the arcing contacts operate in oil, the oil must be changed periodically because of the build up of carbon and other byproducts that are produced by the frequent arcing that takes place.

European transformer manufacturers adopted the resistance-type load tap changer. By the turn of the twenty-first century, the resistance-type load tap changer became the worldwide standard mainly because large power transformers, where LTCs are commonly installed, are no longer built in the United States. One resistance-type load tap changer design is illustrated in Figure 6.15. The transfer switch is open only when the tap changer is in the bridged position, as shown in the figure, with the resistor limiting the circulating current that flows between the taps. Unfortunately, the resistor is also in the load current path, so I^2R heating in the resistor is of concern. Unlike the preventative autotransformer which splits the output voltage between the moving contacts in the bridged position, the output voltage of a resistance-type load tap

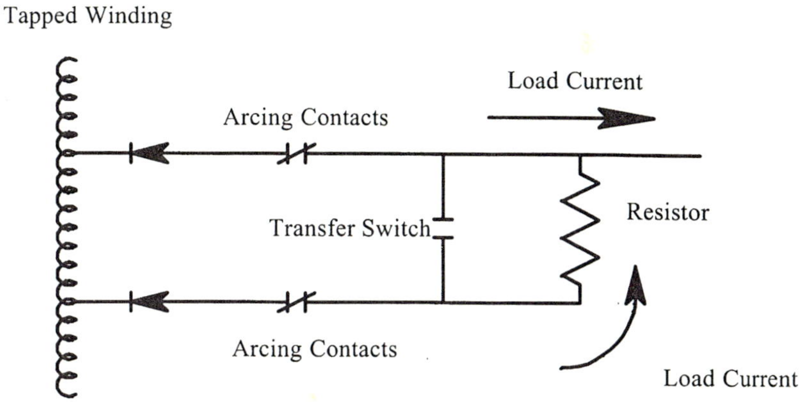

FIGURE 6.15 A resistance-type load tap changer (LTC) momentarily in the bridging position. The resistor restrains short-circuit currents between adjacent taps.

changer is always equal to the voltage of the upper moving contact. Since there are no half-step output voltages available in the bridging positions, there is no advantage in leaving the tap changer in this position. Therefore, the tap changer stays in the bridging position only long enough to let the second moving contact to meet up with the first, thus limiting the I^2R heating in the resistor. Two sets of arcing contacts are required, one for each moving contact. The single transfer switch breaks only the parallel current with both sets of arcing contacts closed, so there is minimal arcing across the transfer switch.

The complete resistance-type tap-changing mechanism, complete with moving contacts, arcing contacts, transfer switch, and resistor can be housed in a single unit that can be easily replaced in the field. This is one of the major advantages of this type of load tap changer. The tap-changing unit is usually located in the main transformer tank with the parts insulated by the same transformer oil that bathes the core and coils, with the exception of the arcing contacts which are in a sealed self-contained oil compartment to prevent contaminating the main transformer with arcing products. Resistance-type load tap changers have a very long life and can perform hundreds of thousands of operations between overhauls.

6.7 VOLTAGE REGULATING TRANSFORMERS

There is a special type of transformer that is designed specifically to regulate line voltage over a relatively narrow output range, typically around ±10% of the incoming voltage. Regulating transformers are technically not power transformers, and many of the industry standards treat the two as different types of transformer. Since the voltage ratio is close to 1.0, it makes sense to use an autotransformer connection and voltage regulators indeed almost exclusively use the autotransformer connection. The series winding is fully tapped end-to-end and it can be connected in the buck as well as the boost mode. In the *neutral* tap position (neither boost nor buck), the series winding is completely disconnected and the regulator does not transform any energy through its windings. A load tap changer is used to change taps of the series winding and a reversing switch is provided to convert the regulator from the buck to boost.

Since the voltage ratio is so close to unity, the multiplication factor F_c of most regulators is quite high and their physical sizes are small compared to the power throughput that they can handle. Of course, since the voltage ratio is close to unity, the effects of short circuits and line surges, which are exaggerated in autotransformers, are even more acute with voltage regulators.

Three-phase voltage regulators are used in substations to regulate the voltages of feeder circuits. Single-phase and three-phase pole-mounted voltage regulators are found in distribution circuits, along lines where voltage sag is a problem.

Maintenance of voltage regulators follows the same guidelines as LTC maintenance. The mechanical drive, controls, main contacts, and arcing contacts all require attention.

6.8 TAP CHANGER AUTOMATIC CONTROLS

The tap-changing mechanism is usually motor-driven and can be controlled manually and automatically. In the automatic mode, the output voltage of the transformer is compared to a reference voltage and a raise/lower signal is sent to the tap changer motor when the output voltage falls outside a specified band, called a *dead band*. The dead band must not be smaller than the voltage between taps; otherwise, it will "hunt" endlessly and burn up the tap changer. The dead band must not be too wide, however, because the purpose of voltage regulation will be defeated. Ordinarily, the dead band is set to a voltage between two and three tap increments. With the tap voltage typically around 1% of the nominal secondary voltage, this provides regulation within a 2–3% range.

If two or more transformers with load tap changers are connected in parallel, then it is important that all transformers operate at the same transformer turns ratio; otherwise, excessive circulating KVAR results. The way this is implemented is to have the tap changer on one of the parallel transformers control voltage as the *lead* tap changer, with the other tap changers as the *followers*. Circuitry is installed on the followers to sense the direction of reactive power flow through each transformer. If too much reactive power is flowing from the primary to the secondary, then its secondary taps are lowered. If too much reactive power is flowing from the secondary to the primary, then secondary taps are raised. Appropriate dead bands are established to prevent the LTCs from hunting and to limit interactions among the tap changer controls.

One such control scheme, called a *load-balancing method*, is depicted in Figure 6.16 for three transformers with voltage regulators supplying a common load bus. If the three transformer impedances equal and if the transform-

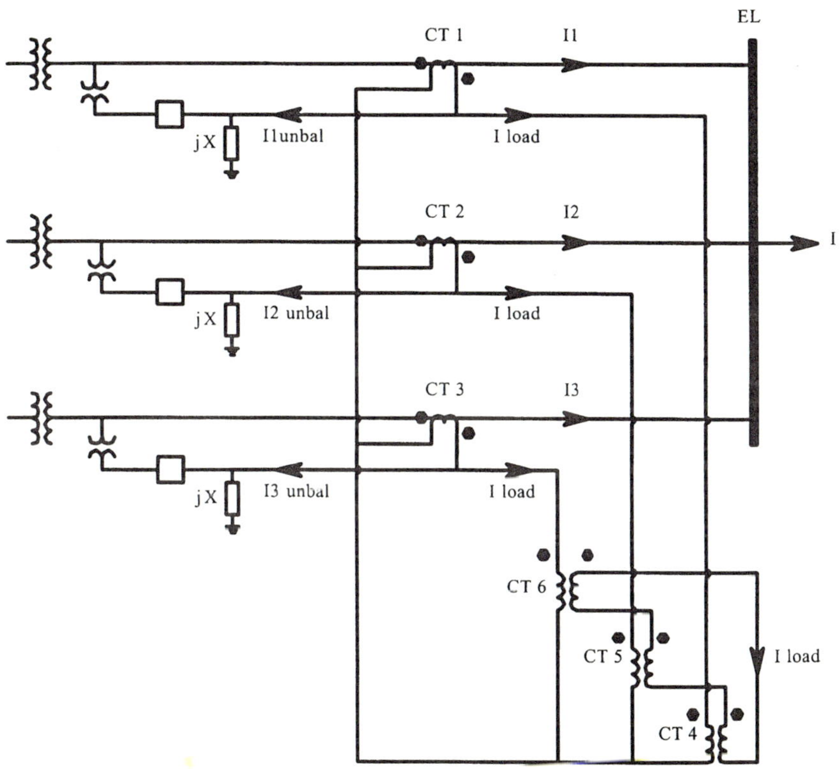

FIGURE 6.16 A load-balancing control scheme for three parallel transformers with load tap changers.

ers are set on the proper tap positions, the transformer secondary currents will all be in phase with the load current and there will be no current unbalance. If one or more transformer is set on the wrong tap, circulating currents will flow in all three transformers. The principle of operation of the load-balancing method is to separate each of the transformer secondary currents into a load-current component and a circulating-current component. The transformer secondary currents flow through current transformers, labeled CT 1, CT 2, and CT 3 in Figure 6.16. The currents at the CT secondaries split into two paths at each of the CT secondary windings.

Path 1 (to the right) goes through a set of auxiliary transformers, labeled CT 4, CT 5, and CT 6. The secondaries of the auxiliary CTs are connected in series, forcing the currents in all three primary windings equal one another, each being one-third of the total load current. By default, the unbalance-current components must flow in path 2 to the left.

Each of the circulating-current components (also called unbalanced-current components) flows through an inductive reactance element, called a paralleling reactor. The paralleling reactors are labeled jX in Figure 6.16. In general, the unbalance-current components of the three transformers are unequal. The voltages developed across the paralleling reactors are added to the *sensed voltages* at the secondary windings of the main transformers, which are used to control the movement of the tap changers.

If transformer 1 is on a higher tap position than transformer 2 or transformer 3, the unbalanced currents flowing through the parallel reactors increase the sensed voltage at transformer 1 and reduce the sensed voltages at transformers 2 and 3. This causes transformer 1 to lower its taps and transformers 2 and 3 to raise their taps. If transformer 1 is on a lower tap position than transformers 2 and 3, the unbalanced currents flowing through the parallel reactors decrease the sensed voltage at transformer 1 and increase the sensed voltages at transformers 2 and 3. This causes transformer 1 to raise its taps and transformers 2 and 3 to lower their taps.

6.9 VARIABLE PHASE SHIFTING TRANSFORMERS

If two areas are connected with a number of transmission paths, it is sometimes necessary to control the flow of real power through the transmission network. Phase-shifting transformers are installed to force a phase angle shift between the ends of a transmission line and create a real power flow. A phase shift of a few degrees is all that's required to move massive amounts of power around the system, so a phase-shifting transformer only needs to transform a tiny fraction of the power that is moving through the system. The power flow between two buses can be approximated by the following equation.

$$P = \frac{V_S \times V_R \times \sin\theta}{X} \tag{6.9.1}$$

where

P = real power flow in per unit
V_S = per unit voltage of the sending bus
V_R = per unit voltage of the receiving bus (assumed to lag V_S)
θ = phase angle difference between V_S and V_R
X = per unit reactance between the sending and receiving bus

The situation is analogous to a voltage-regulating transformer, which can create a flow of reactive power by a slight change in voltages. Again, an autotransformer is the logical choice, using the special phase-shifting connection shown in Figure 6.17. The common tapped series windings are equivalent

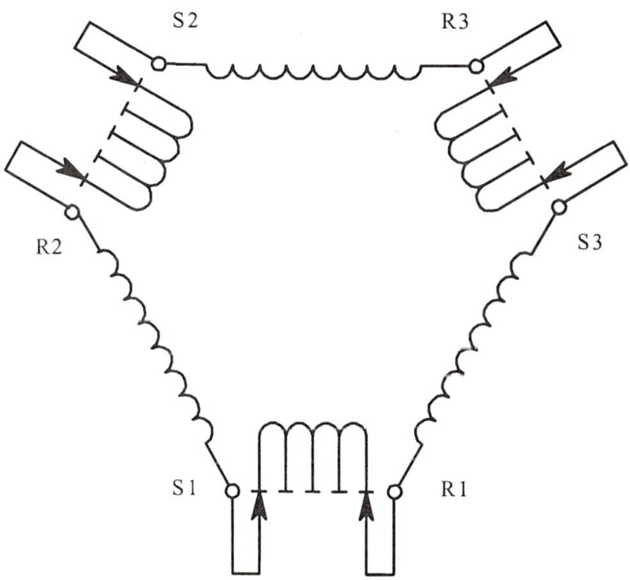

FIGURE 6.17 The winding configuration of a variable phase-shifting transformer used to control real power flow. Note that the voltages of the sending terminals and receiving terminals have equal magnitudes.

to the common and tapped series windings in a voltage-regulating transformer, except they are interconnected differently to produce a phase angle shift instead of a voltage change. There are six load tap changers in this configuration although some designs use three load tap changers plus a reversing switch instead.

The voltages at the sending terminals (S1-S2-S3) are shown lagging the voltages at the receiving terminals (R1-R2-R3), but since each tap changer can travel the full range of taps, the phase shift can also be reversed with the sending terminals leading the receiving terminals. Note that the magnitudes of the line-to-line voltages at the sending and receiving terminals are equal to each other. Therefore, the change in voltage is a pure phase shift.

If reactive power flow needs to be regulated with real power flow regulated independently, voltage and phase angle control can be combined within a single regulating transformer with two sets of tap changers operating independently. One set of tap changers regulates the in-phase voltages and the other set of tap changers regulates the quadrature-phase voltages, which are combined at the output. A typical phase shifting transformers may be capable of a variable phase angle shift up to 20°; however, if X is on the order of a few percent, a fairly small phase angle shift may be sufficient for diverting large blocks of power.

Example 6.1

Suppose a phase-shifting transformer, such as the one in Figure 6.17, is connected to a 500 kV system. The phase shifter regulates 1000 MW of power from the receiving terminals to the sending terminals by shifting the phase angle of the sending terminals by −3°. How much MVA is actually transformed by the phase shifter under these conditions?

From the voltage vectors of the sending and receiving voltages, the voltages across the tapped windings E_T and the voltages across the common windings E_C are calculated:

$$E_T = 2 \times \frac{500{,}000 \text{ V}}{1.732} \times \sin\frac{3°}{2} = 15{,}113 \text{ V}$$

$$E_C = 2 \times \frac{500{,}000 \text{ V}}{1.732} \times \sin\frac{117°}{2} = 492{,}272 \text{ V}$$

Let I_T = current flowing in the tapped windings and I_C = current flowing in the common windings.

$$I_C = I_T \times \frac{15{,}113}{492{,}272} = 0.0307 \times I_C$$

The current out of S1 based on 1000 MW of flow:

$$|I_{S1}| = \frac{1{,}000{,}000{,}000 \text{ W}}{1.732 \times 500{,}000 \text{ V}} = \frac{1155 \text{ A}}{\text{phase}}$$

From the vector diagram of the transformer, the currents in all windings flow in a circular loop, but the phase angles of the currents in the tapped windings and the two adjacent common windings are 120° apart. The current flowing out of terminal S1 are the sum of the winding currents at terminal S1:

$$|\mathrm{I}_{S1}| = |I_T\angle 0° - I_C\angle 120°| = I_T \times |1\angle 0° - 0.0307\angle 120°|$$
$$= 1.016 \times I_T = \frac{1155 \text{ A}}{\text{phase}}$$

Solving for I_T:

$$I_T = \frac{1155 \text{ A}}{1.016} = 1137 \text{ A}$$

The MVA transformed is equal to the winding currents times the winding voltages:

$$\text{MVA} = \frac{15{,}113 \text{ V} \times 1{,}137 \text{ A}}{1{,}000{,}000} = 17.2 \text{ MVA per phase}$$
$$= 3 \times 17.7 \text{ MVA per phase} = 51.6 \text{ MVA total}$$

The total transformed MVA slightly greater than 5% of the total MVA flowing through the phase shifter. Like a voltage regulator, the core and coils of the phase shifter can be surprisingly small for the huge amount of power flow it is able to regulate.

Example 6.2

Two identical transformers, each having a j0.15 per unit impedance, supply a substation load. Their load tap changer controls are operating in a load-balancing scheme. Transformer 1 is set on the 1.05 per unit secondary load tap, and transformer 2 is set on the 1.00 per unit secondary load tap position, with unity voltages assumed at the primaries of both transformers. The load may be represented as a constant impedance of $1.0\angle 20°$ per unit. The paralleling reactors each have a 1.0 impedance.

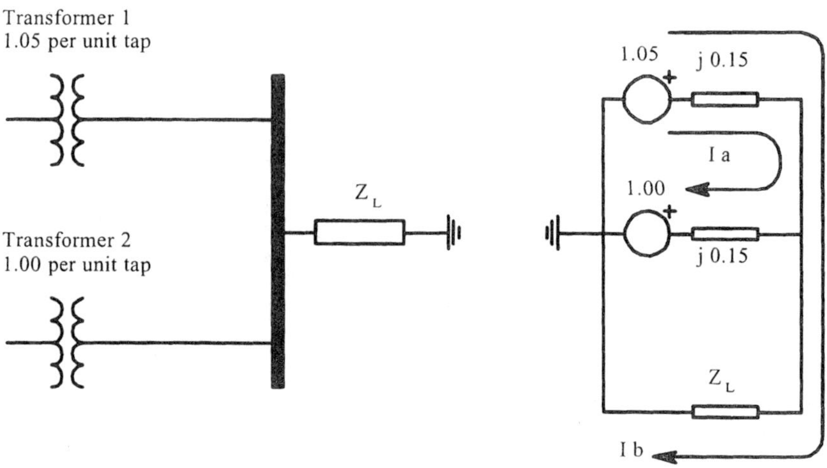

FIGURE 6.18 Circuit diagrams for Example 6.2.

1. Find the voltage at the load and the currents at the secondaries of the two transformers and sketch these quantities as vectors.
2. Decompose each of the secondary currents into a load-current component and an unbalanced-current component.
3. Find the voltages developed across the paralleling reactors and sketch these voltages along with the load voltage.

The circuit is shown along with its equivalent circuit in Figure 6.18. The load impedance is converted into rectangular form:

$$1.0\angle 20^\circ = 0.94 + j0.342 \text{ per unit}$$

The loop equations are as follows:

$$I_a \times j0.30 + I_b \times j0.15 = 1.05\angle 0^\circ - 1.00\angle 0^\circ \text{ per unit}$$
$$I_a \times j0.15 + I_b \times (0.94 + j0.342) = 1.05\angle 0^\circ \text{ per unit}$$

The solutions of the loop currents are as follows:

$$I_a = 0.50505\angle 182.7^\circ = -5.04508 - j0.02336 \text{ per unit}$$
$$I_b = 1.04893\angle 344.1^\circ = 1.00901 - j0.2866 \text{ per unit}$$

The load voltage is calculated:

$$E_L = I_b \times 1.0\angle 20^\circ = 1.04893\angle 4.1^\circ \text{ per unit}$$

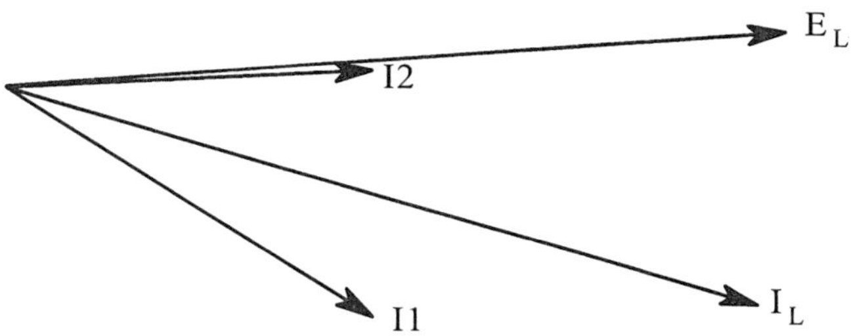

FIGURE 6.19 Voltage and current vectors for Example 6.2.

The transformer secondary currents are calculated:

$$I1 = I_a + I_b = 0.5045 - j0.310 = 0.592\angle{-31.6°} \text{ per unit}$$
$$I2 = -I_a = 0.50505\angle 2.7° = 0.5045 + j0.0238 \text{ per unit}$$

The load voltage, load current, and transformer currents are plotted in Figure 6.19. In the load-balancing scheme, the load-current components of the two transformers are assumed to be equal to each other and comprise the total load current when they are added together. Therefore, each of the load-current components is equal to one-half of the load current:

$$I_{L1} = I_{L2} = {}^1\!/_2 I_b = 0.5245\angle 344.1° = 0.5045 - j0.1433 \text{ per unit}$$

The unbalance-current components are equal to the transformer secondary currents minus the load-current components:

$$I_{U1} = I1 - I_{L1} = (0.5045 - j0.310) - (0.5045 - j0.1433)$$
$$= -j0.1667 \text{ per unit}$$
$$I_{U2} = I2 - I_{L2} = (0.5045 + j0.0238) - (0.5045 - j0.1433)$$
$$= j0.1195 \text{ per unit}$$

The voltages developed across the paralleling reactors are equal to the unbalance-current components times the impedances of the paralleling-reactor, which are equal to j1.0 per unit:

$$E_{U1} = I_{U1} \times j1.0 = 0.1667 \text{ per unit}$$
$$E_{U2} = I_{U2} \times j1.0 = -0.1195 \text{ per unit}$$

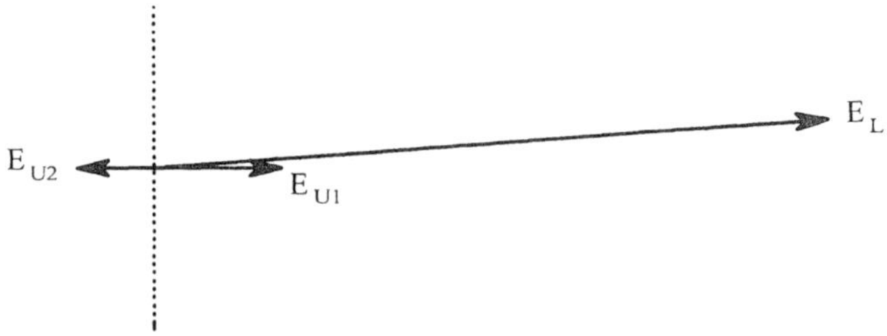

FIGURE 6.20 Voltage vectors for the load bus voltage and voltages developed across the paralleling reactors in Example 6.2.

The load voltage and the voltages developed across the paralleling reactors are plotted in Figure 6.20. Note that when E_{U1} and E_{U2} are added to the sensed voltage E_L, the resulting voltages tend to cause transformer 1 to lower taps and transformer 2 to raise taps.

Example 6.3

GenCo Unit 2 is a generating plant with a net output of 1100 MW, capable of a 90% lagging power factor. The generator step up transformer for the unit has three 405 MVA FOA single-phase transformers connected in a Δ-Grd.Y bank. Each of these transformers has a total of eight coolers. The transformer test report is shown in Figure 3.16 in Chapter 3 of this book.

One summer afternoon with Unit 2 running at full output, a generator step-up transformer trouble alarm was received in the Unit 2 control room. An electrical maintenance supervisor was dispatched to the transformers and he observed that transformer serial number C-05213-5-1 had a high-temperature alarm, which was confirmed by the indications of the top oil temperature gauge and the winding temperature gauge. The supervisor also noticed that half of the cooling fans on that transformer were not running. The circuit breaker supplying these fans had tripped, and there was evidence of an electrical fault in the control cubicle, severely damaging some relays and wiring. It was estimated that repairs would take at least 12 h, provided a spare relay could be found. Meanwhile, the transformer temperature continued to increase.

The plant superintendent had to choose whether to shut down the unit or reduce the net output. Based on the transformer test report, how much net output can the transformer bank safely handle with one of the transformers in the degraded condition described above?

These transformers are cooling class FOA, which means they have no self-cooled rating. With half of the cooling fans inoperative on Serial Number C-05213-5-1, it can be assumed that half of the cooling capability of this transformer is lost. Since the transformers are banked, the load on all three phases must be reduced to the point where the load on Serial Number C-05213-5-1 is within its thermal capability.

In order to provide a margin of safety, it was decided not to take credit for any thermal capability beyond nameplate ratings. Therefore, the new transformer loading is to be based on derating the nameplate MVA load of the affected unit.

From the test report shown in Figure 3.16, the total losses consist of 280.5 kW in core losses and 909.9 kW in copper losses at 405 MVA load, for a total of 1190.4 kW in losses. These losses must be reduced by 50% with half of the coolers inoperative.

$$\text{New total losses} = 0.5 \times 1190.4\ \text{kW} = 595.2\ \text{kW}$$

Since the core losses are not a function of load, these are constant. Therefore, the new copper losses are calculated as follows:

$$\begin{aligned}\text{New copper losses} &= \text{new total losses} - \text{core losses}\\ &= 595.2\ \text{kW} - 280.5\ \text{kW} = 314.7\ \text{kW}\end{aligned}$$

Copper losses are proportional to the load current squared. The ratio of the new copper losses to the nameplate copper losses is equal to the ratio of the new load current to the nameplate load current.

$$\frac{(\text{New load current})^2}{(\text{Nameplate load current})^2} = \frac{314.7\ \text{kW}}{909.9\ \text{kW}} = 0.346$$

The load current is approximately equal to the MVA load. Therefore, the new MVA load, based on derating the nameplate MVA load, is found by the following relationship:

$$\begin{aligned}\text{New MVA load} &= \text{nameplate MVA load} \times 0.346^{1/2}\\ &= 405\ \text{MVA} \times 0.588 = 238\ \text{MVA}\end{aligned}$$

The load limit of the bank is equal to 3 times the load limit of the affected transformer. This establishes the net generator output.

$$\text{Net generator MVA output} = 3 \times 238\ \text{MVA} = 714\ \text{MVA}$$

At a 90% lagging power factor, the net MW output is 90% of the net generator MVA output:

$$\text{Net generator MW output} = 0.9 \times 714\ \text{MVA} = 642.6\ \text{MW}$$

This reduces the generator output by approximately 42%, which is still preferable to shutting it down.

REFERENCE

1. IEEE Std. C57.12.00–2000. IEEE Standard General Requirements for Liquid-Immersed Distribution, Power, and Regulating Transformers. New York: Institute of Electrical and Electronics Engineers, Inc., New York 2000, Page 9.

7

Reading and Applying Nameplate Information

Every distribution and power transformer has a metal nameplate attached to the tank that gives vital information on how the transformer is to be connected and operated. The information is printed or stamped on the nameplate so it is a permanent part of the transformer. A transformer's nameplate has been compared to a birth certificate because it contains so many vital statistics that will follow it throughout its service life. In this chapter, examples of actual transformer nameplate information are illustrated and the data provided on the nameplate are discussed.

7.1 MINIMUM NAMEPLATE REQUIREMENTS

The minimum information to be shown on a transformer nameplate depends on the KVA rating of the transformer as specified in the standards. The standards require the following information for transformers rated above 500 KVA [1].

- Name of manufacturer
- Serial number
- Month/year of manufacture

- Cooling class. Where the class of transformer involves more than one rating, all ratings are shown. Windings having different ratings have their individual KVA ratings described. If the transformer has more than one temperature rating, these ratings are shown on the nameplate. Provisions for future cooling equipment are indicated.
- Number of phases
- Frequency
- KVA or MVA rating
- Voltage ratings. The voltage ratings of a transformer or autotransformer windings are separated by a dash (-).
- Tap voltages. The tap voltages of an individual winding are separated by a slash (/).
- Rated temperature rise, °C
- Polarity (single-phase transformers)
- Phasor or vector diagram (polyphase transformers)
- Percent impedance. The percent impedance is specified between each pair of windings with the voltage connection and the KVA base stated. The percent impedance shall be tested.
- Basic lightning impulse insulation levels (BIL). The BIL of each winding and each bushing are specified.
- Approximate mass of the core and coils, tank and fittings, insulating oil, total weight, and heaviest piece.
- Connection diagram. The connection diagram shows all winding terminations with a schematic plan view showing all fixed accessories. The schematic plan view generally shows the low-voltage side of the transformer at the bottom . The location and polarity shall be shown for all current transformers used for metering, relaying, or drop compensation.
- Installation and operating instructions references
- The word ''transformer'' or ''autotransformer''
- Suitability for step-up operation. Since the MVA rating is generally specified at the output winding at the nominal output winding voltage, the MVA rating for step-up operation will generally be different than the MVA rating for step-down operation, unless specified.
- Maximum positive and negative operating pressures of the oil preservation system, kPa or psi
- Maximum negative pressure of the tank for vacuum filling
- Liquid level below the top surface of the highest point of the highest manhole flange at 25°C
- Change in liquid level per 10°C change in liquid level

- Oil volume of each transformer compartment
- Type of insulating liquid. The nameplate shall also have the following note: "Contains no detectable level of PCB (less than 2 PPM) at the time of manufacture."
- Conductor material of each winding

7.2 MANUFACTURER'S INFORMATION

The manufacturer's name usually appears prominently at the top of the nameplate. Just below the manufacturer's name is a unique serial number or shop order number to identify the particular unit so its manufacturing and design records can be traced. In some respects, the nameplate is equivalent to a birth certificate.

7.3 COOLING CLASS, NUMBER OF PHASES, AND OPERATING FREQUENCY

The transformer cooling class, number of phases, and the operating frequency are usually specified near serial number. As we saw in Chapter 6, there are a variety of cooling classes. The more common types are reviewed in Table 6.1.

The second and third cooling stages provide additional loading capabilities with respect to the OA rating. See Table 7.1.

TABLE 7.1 Nominal Ratings of Various Cooling Classes as a Percentage of the Self-Cooled Rating (per NEMA)

Transformer OA rating		Cooling class	Top rating as a percentage of OA rating
Single phase	501–2,499 KVA		
Three phase	501–2,499 KVA	OA/FA	115%
Single phase	2,500–9,999 KVA		
Three phase	2,500–11,999 KVA	OA/FA	125%
		OA/FA	
		OA/FOA	$133\frac{1}{3}\%$
Single phase	≥10,000 KVA	OA/FA/FA	
Three phase	≤12,000 KVA	OA/FA/FOA	$166\frac{2}{3}\%$
		OA/FOA/FOA	

Source: Ref. 2.

7.4 VOLTAGE RATINGS

The voltage rating of the transformer is specified in terms of both the line-to-line system voltages and the winding voltages. For example, a 230 kV Grd.Y to 69 kV Grd.Y transformer with a 12.47 kV tertiary would have the following voltage rating on its nameplate:

230000GR.Y/132800-69000GR.Y/39840-12470

The hyphens separate the voltage ratings of one winding from the voltage ratings of the other windings. In this illustration, 230000GR.Y/132800 is the rating of the high-voltage winding, 69000GR.Y/39840 is the rating of the low-voltage winding, and 12470 is the rating of the tertiary winding. The voltage ratings are always given in volts.

When a slash is shown in the voltage rating of a winding, this indicates that the winding is Y-connected with the phase-to-phase system voltage appearing first followed by the slash and then the winding voltage. The winding voltage is the phase-to-neutral voltage of the system. If a designator GR.Y, GRD.Y, or Grd.Y appears, this indicates that the neutral end of the winding is intended to be grounded, which probably means that there is reduced insulation near the neutral end of the winding. In this case, operating the winding ungrounded may not be safe.

The example 230000GR.Y/132800-69000GR.Y/39840-12470 designates the 12,470 V tertiary winding by a single number without a slash appearing because the winding ratings and the line-to-line voltage of the system are the same. In other words, the windings are connected phase-to-phase in a Δ.

If the nameplate designates a winding voltage is designated with a Y but without a slash, the winding is permanently Y-connected without bringing out a neutral connection. For example, 66000-12470Y shown on the nameplate indicates a Δ-connected primary winding rated 66,000 V with a Y-connected secondary winding suitable for connection to a 12,470 V system.

Example 7.1

Two transformers have the following nameplate voltage ratings:

Transformer 1: 69000-12470GR.Y/7200

Transformer 2: 69000GR.Y/39840-12470

Explain the difference between the two transformers. What are the TTRs of these transformers? Can the two transformers be operated in parallel?

Transformer 1 has a Δ-connected primary with a line-to-line rating and a winding rating of 69,000 V. The secondary is grounded-Y connected with a line-to-line rating of 12,470 V and a winding rating of 7200 V. The TTR of this transformer equals the ratio of the winding ratings: 69,000/7200 = 9.583.

Transformer 2 has a grounded-Y connected primary with a line-to-line rating of 69,000 V and a winding rating of 39,840 V. The secondary is Δ-connected with a line-to-line rating and a winding rating of 12,470 V. The TTR of this transformer equals the ratio of the winding ratings: 39,840/12,470 = 3.195.

Since the TTRs of these transformers are so different, we could be tempted to state that they cannot be operated in parallel. The important consideration is not whether the TTRs are the same, but whether the secondary voltages are matched when the primaries are connected to the same source voltage. If both transformers have the same phase angle displacement, which is −30° for a standard design, and the percent impedance values are nearly the same, these transformers may be operated in parallel.

7.5 KVA OR MVA RATINGS

The nominal KVA or MVA ratings are shown on the nameplate for a specified winding temperature rise and cooling mode. For example, consider an OA/FOA/FOA cooling class transformer with "45/60/75 MVA Continuous at 65°C Temperature Rise" specified on its nameplate. This transformer is designed to operate continuously at 45 MVA as a self-cooled (OA) transformer, or at 60 MVA with one stage of forced oil and forced air cooling, or at 75 MVA with two stages of forced oil and forced air cooling with an average winding conductor temperature at 65°C above the ambient air temperature. Notice that the ratio of the rating with the first stage of forced cooling to the OA rating is 60/45 or 133 1/3% and the ratio of the rating with the second stage of forced cooling to the OA rating is 75/45 or 166 2/3%, which conforms to the NEMA standard, as shown in Table 7.1.

Some oil-filled transformers have KVA and MVA ratings specified at both 65°C rise and 55°C rise. This is because transformers built prior to the 1960s were limited to an average winding temperature rise of 55°C because of the types of insulating materials that were available at that time. Standard ratings developed for these transformers were based on the 55°C temperature rise. Later, improved insulating materials became available, which allowed transformers to operate at a 65°C temperature rise. The KVA ratings at 65°C rise were higher than the "standard" ratings that were previously developed. Therefore, these transformers were built to the "standard" KVA ratings for

CLASS	OA	FOA	FOA	Rise °C	
MVA	45	60	75	65	BURIED TERTIARY 35% CAPACITY

FIGURE 7.1 Part of a transformer's nameplate showing the MVA ratings by cooling class.

55°C temperature rise but could be operated at higher KVA ratings at 65°C temperature rise, so both sets of ratings were shown. Today, most transformer nameplates omit the 55°C ratings entirely and are built for standard KVA ratings at 65°C temperature rise.

Certain high-performance insulating materials such as Nomex and silicone fluids allow operation at 75°C rise or even higher temperatures. The nameplate ratings would reflect the KVA ratings at the higher temperature rise.

For two-winding transformers, there is usually only one set of KVA or MVA values shown, because it is understood that the primary and secondary windings must have the same KVA capabilities. This is not true for three-winding transformers, however, since the tertiary is often not designed to the same KVA capabilities as the primary and secondary windings. The nameplate for a three-winding transformer could read as shown in Figure 7.1. Sometimes, the details of the tertiary ratings are given in the winding connection tables and diagrams, discussed below.

7.6 WINDING CONNECTION DIAGRAM

The actual winding connections are shown in a diagram with each winding and its taps labeled. A set of tables then specifies the voltage ratings, ampere ratings, and the connections for all the available taps. For transformers with load tap changing equipment, the connection diagrams and the accompanying tables are quite extensive.

The connection diagram usually also gives the general physical layout of the transformer, showing the placement of the bushings and the locations of current transformers (CTs) and a schematic representation of the load tap changing equipment, including the preventative autotransformer, moving contacts, arcing contacts, transfer switch, and reversing switch.

A portion of an actual nameplate that shows the winding connection diagram is illustrated in Figure 7.2. The nameplate depicted is rather interesting. The transformer has a load tap changer. From the connection diagram we

3 PHASE	60 HZ	CLASS	OA	FOA	FOA	TEMP RISE C	H.V. 225000GRD.Y / 129900	L.V. 13800GRD.Y / 7970	15 % IMPEDANCE
S/N 70162 - 1		MVA	45	60	75	65	BURIED TERTIARY 35% CAPACITY		AT 45 MVA

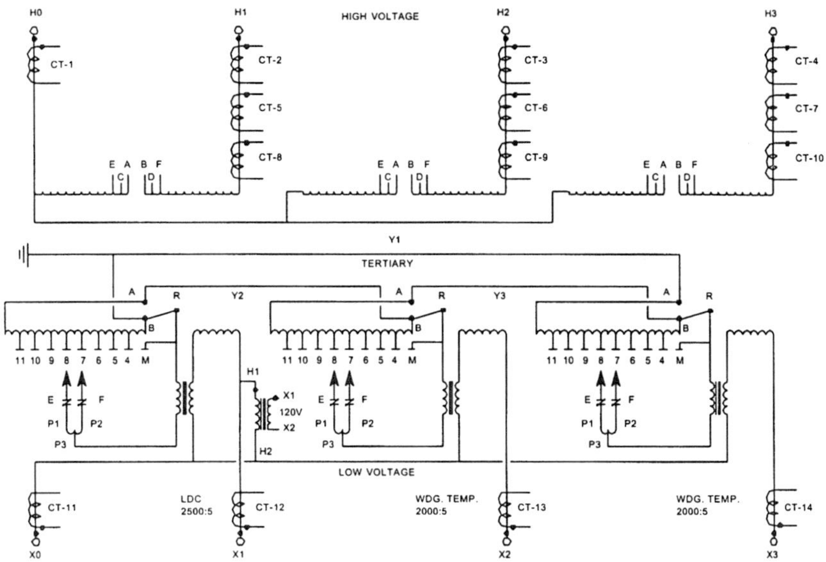

CAUTION: TAP CHANGER FOR DE-ENERGIZED OPERATION AND MUST NOT BE OPERATED WHEN TRANSFORMER IS ENERGIZED

VOLTS	AMPS AT 75C	TAP	CONNECTS
236250	1833	1	A - B
230625	1878	2	B - C
225000	1925	3	C - D
219375	1974	4	D - E
213750	2026	5	E - F

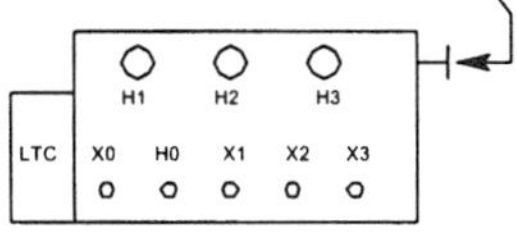

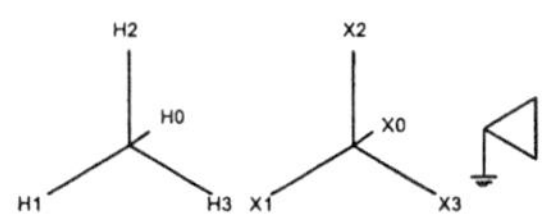

RAISE REVERSING SWITCH R TO B			
LOW VOLTAGE	AMP AT 75C	IND. POS.	CONTACTS ON
15870	2720	16	11 - 11
15741	2751	15	11 - 10
15611	2774	14	10 - 10
15482	2797	13	10 - 9
15353	2720	12	9 - 9
15222	2844	11	9 8
15094	2869	10	8 - 8
14964	2894	9	8 - 7
14835	2919	8	7 - 7
14706	2945	7	7 - 6
14576	2971	6	6 - 6
14447	2997	5	6 - 5
14318	3024	4	5 - 5
14188	3052	3	5 - 4
13929	3080	2	4 - 4
14964	3109	1	4 - M
13800	3138	N	M

LOWER REVERSING SWITCH R TO A			
LOW VOLTAGE	AMP AT 75C	IND. POS.	CONTACTS ON
13800	3138	N	M
13671	3168	1	M 11
13541	3198	2	11 - 11
13412	3229	3	11 - 10
13283	3260	4	10 - 10
10150	3292	5	10 - 9
13024	3325	6	9 - 9
12894	3358	7	9 - 8
12765	3392	8	8 8
12636	3427	9	8 - 7
12506	3462	10	7 7
12377	3499	11	7 - 6
12478	3536	12	6 - 6
12118	3573	13	6 - 5
11989	3612	14	5 - 5
11859	3651	15	5 - 4
11730	3692	16	4 - 4

FULL WAVE IMPULSE LEVEL KV	
HV WINDING	750
LV WINDING	110
HV WDG. NEUTRAL	110
HV NEUT. BUSHING	150
LV NEUT. BUSHING	150
HV BUSHING	900
LV BUSHING	150

FIGURE 7.2 Part of a transformer's nameplate showing the voltage ratings, MVA ratings, percent impedances, connection diagram, physical layout, vector diagram, tap connections, CT connections, and BIL ratings.

see that the buried tertiary is also a tapped winding that supplies a buck/boost voltage to the secondary windings through auxiliary transformers connected between the tertiary and the secondary. Therefore, the tertiary simultaneously provides four important functions:

1. It provides a path for third harmonic currents.
2. It helps stabilize voltages in the Y-Y primary-secondary connection.
3. It provides a grounding bank action by providing a path for zero-sequence currents.
4. It provides the necessary voltage taps for regulating the low-side voltage.

The only function that the buried tertiary cannot perform is to supply an external load. The voltage rating of the buried tertiary is not given because it cannot be connected to a system voltage, but one corner of the Δ connection is grounded internally. This grounding is done so the winding potential voltage does not "float" because of capacitive coupling to the other windings. Without this ground connection, capacitively induced voltages are indeterminate and could be large enough to cause insulation damage.

The voltage taps for the primary and secondary are shown on the connection diagram and on the winding rating tables in Figure 7.2. These also specify which terminal numbers and letters are connected for each tap. This transformer has a total of 14 current transformers that are used for metering, protective relaying, and other purposes. Note the CTs marked "LDC" and "WDG. TEMP." The term LDC stands for *line drop compensation*. The LDC CT supplies metered line current to a compensating device in the voltage regulator controls. The compensating device effectively moves the voltage control point into the system connected to the secondary winding. The CT labeled WDG.TEMP supplies current to the winding temperature gauges. As previously discussed, these gauges use a heating element surrounding a temperature probe mounted in the top oil in order to mimic the winding temperature. The ratios of these CTs would be shown on an actual nameplate, but this information is not shown in Figure 7.2.

Just below the connection diagram is a layout sketch showing the physical locations of the bushings, the load tap changing compartment and the operating handle for the tap changer at deenergized conditions.

The load tap changer is represented schematically in the connection diagram. Note the terminals labeled P1, P2, and P3. These terminals correspond to the connections to the preventative autotransformer. The two series arcing contacts per phase that are in series with the movable contacts are shown as well.

7.7 PHASOR OR VECTOR DIAGRAM

All nameplates for polyphase transformers include vector diagrams to show the phase relationships between the windings. If the transformer has a nonstandard phase shift, this will be indicated on the vector diagram. Some transformers have terminal boards that can be wired to provide different configurations such as either Δ-Y or Y-Y, or unusual phase displacements, such as Y-Y with a 180° phase displacement between the primary and secondary. Each of the possible configurations is shown on the vector diagram.

The vector diagram on the nameplate shown in Figure 7.2 shows the standard phase angle displacement of 0° for a Y-Y connected transformer.

7.8 WEIGHTS AND OIL CAPACITY

To avoid mechanically overloading equipment when handling and rigging large transformers, weight is an important consideration. Also, if insulating oil is to be removed for shipping, it is important to know quantities of oil that are involved. The nameplate provides a convenient reference for these data, which is permanently attached to the transformer. The nameplate has a table of approximate weights (that are usually somewhat conservative) and the oil capacity, usually given in U.S. gallons.

The weights and oil volumes are broken down by components and oil compartments. The untanking weight is the weight of the heaviest piece that must be lifted for disassembly. There is also information showing the internal oil level at 25°C that is used to properly fill the transformer, plus an indication of the change in oil level per change in temperature. Figure 7.3 shows the manner in which this information is typically shown on actual nameplates.

Example 7.2

A transformer is to be loaded onto a rail car after the oil is drained from the main tank and the LTC compartment. A 150 T crane is available. Using the information in Figure 7.3, is the crane's capacity adequate?

The untanking weight of the unit (core and coils) is 118,000 lb and the weight of the tank and fittings is 88,000 lb. Thus, the net weight of the transformer without oil is 118,000 lb + 88,000 lb = 206,000 lb = 103 T. The capacity of the crane is adequate to lift the transformer without oil. Note that when the transformer is filled with oil, the total weight is 306,200 lb = 153.1 T, which exceeds the capacity of the crane.

APPROXIMATE WEIGHT	LBS.
UNTANKING	118000
TANK & FITTINGS	88000
OIL MAIN TANK 13000 GALS	97500
OIL LTC COMPARTMENT 360 GALS	2700
TOTAL WEIGHT	306200

LIQUID LEVEL BELOW TOP OF MANHOLE AT 25C [12.6] INCHES. LIQUID LEVEL GANGES [1.17] INCHES PER 10C CHANGE IN LIQUID TEMP. OPERATING PRESSURE RANGE OF OIL PERSERVATION SYSTEM [0.5] P.S.I. POSITIVE TO [8] POSITIVE. TANK DESIGNED FOR [15] P.S.I. VACUUM FILLING. TRANSFORMER IS FILLED WITH INHIBITED OIL.

FIGURE 7.3 Part of a transformer's nameplate showing weights, oil capacities, oil level, and operating pressure range.

Example 7.3

The oil level of a particular transformer is shown in Figure 7.3. The transformer is being filled with oil at a substation at an ambient temperature of 30°C (86°F). Where should the oil level be after when the filling is complete?

According to Figure 7.3, the oil level at 25°C is 12.6 in. below the top of the manhole. The change in oil level is 1.17 in. per 10°C change in oil temperature. Since the ambient temperature is 5°C the reference temperature, the change in oil level is one-half of 1.17 in., or 0.585 in. above the reference point of 12.6 in. Therefore, the proper oil level is 12.6 in − 0.585 in. = 12.015 in. below the top of the manhole.

7.9 OPERATING PRESSURE RANGE

Most modern power transformers are designed to be "vacuum oil filled." Vacuum oil filling is a process that will be covered in more detail in Chapter 8. Before attempting to "pull vacuum" on a transformer, it is important to consult the nameplate first to verify that the tank is designed to withstand the negative pressures. The positive range of operating pressures is also specified on the nameplate.

The nominal atmospheric pressure is 14.7 psi at sea level. Thus, when the transformer contains a perfect vacuum, the negative pressure on the tank will be approximately 15 psi at sea level, and slightly less at higher elevations. In the example shown in Figure 7.3, we see that this transformer is designed to be vacuum oil filled under a "perfect vacuum" equal to about 15 psi of negative pressure.

7.10 IMPEDANCE

The transformer impedance, sometimes referred to as "impedance volts" on the nameplate, is specified for each of pair of windings as a percent value on a specific KVA base. For two-winding transformers, only one value of impedance is given, corresponding to the high-to-low voltage series impedance at a KVA base that usually corresponds to the KVA rating of the primary and secondary windings. For a transformer having multiple cooling classes, the KVA base is usually specified as the OA rating of the transformer.

For a transformer having three or more windings, the situation is complicated by the fact that not all of the windings have the same KVA rating. Therefore, the KVA base used to for the percent impedance of each pair of windings must be specified on the nameplate.

For example, a three-winding transformer may have its impedances shown on its nameplate as shown in Figure 7.4. Note that the percent impedance between the primary and secondary windings is on a higher KVA base than the percent impedances given for the primary to tertiary and the secondary to tertiary. When doing analysis using the nameplate impedance values, care must be exercised to convert the percent impedances to a common KVA base.

Since the nameplate is normally fabricated before transformer assembly is completed, some of design data is etched on the nameplate. The last two items to be determined are the transformer serial number and the percent impedance values that must be determined by test. Therefore, the percent impedances are usually stamped on the nameplate just before the transformer is shipped from the factory.

IMPEDANCE VOLTS 16 PERCENT 230000GR.Y-69000GR.Y VOLTS AT 75000 KVA

IMPEDANCE VOLTS 14 PERCENT 230000GR.Y-12470 VOLTS AT 26250 KVA

IMPEDANCE VOLTS 12 PERCENT 69000GR.Y-12470 VOLTS AT 26250 KVA

FIGURE 7.4 Nameplate designation of percent impedances of a three-winding transformer.

Example 7.4

Convert the percent impedance values shown in Figure 7.4 into per unit impedances on a common 100 MVA base.

Primary to secondary:	0.16×100 MVA/75 MVA $= 0.213$ per unit
Primary to tertiary:	0.14×100 MVA/26.26 MVA $= 0.533$ per unit
Secondary to tertiary:	0.12×100 MVA/26.26 MVA $= 0.457$ per unit

7.11 BASIC INSULATION LEVEL

The basic insulation levels (BILs) of the windings at each of the terminals and the BIL of the bushings are shown on the nameplate. These are given as full-wave test values in kV.

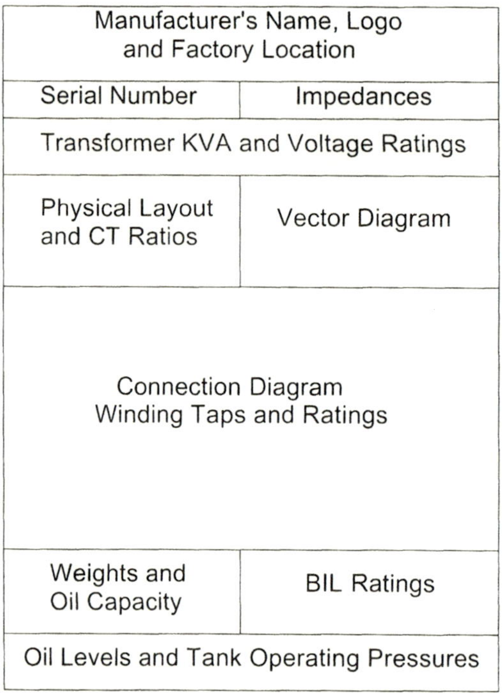

FIGURE 7.5 Overall layout of the complete transformer nameplate.

7.12 NAMEPLATE LAYOUT

There is some flexibility in the standards as to how the nameplate is to be laid out. Each manufacturer arranges the data somewhat differently. Some customers specify a particular layout that is suitable to their needs. Figure 7.5 shows one typical nameplate layout as a block diagram.

REFERENCES

1. IEEE Std. C57.12.00-2000. IEEE Standard General Requirements for Liquid-Immersed Distribution, Power, and Regulating Transformers. New York: Institute of Electrical and Electronics Engineers, Inc., New York, 2000, pp. 24–26.
2. NEMA Standards Publication No. TR 1–1980. Transformers, Regulators and Reactors. Washington, DC: National Electric Manufacturers Association, Part 0, p. 1.

8

Maintenance, Testing, Troubleshooting, and Reliability

Now that we are familiar with the fundamentals of transformer theory, design, and operation, these principles can be applied to maintenance, testing, and troubleshooting. A transformer may have a very long life, or it may fail minutes after first being energized.

The life of an average transformer usually depends to some extent on how well it is treated. Since large power transformers can cost several millions of dollars, performing regular transformer maintenance is a very cost-effective practice.

8.1 GOOD UTILITY PRACTICES

Utility companies have enormous amounts of money invested in transformers of all types, including distribution and power transformers. Utilities also have a long history and have developed methods, procedures, and philosophies that over time have proven very effective in prolonging equipment life. These are collectively referred to as *good utility practices* and it is instructive to review these practices as applied to power transformers.

One utility has an instruction [1] that covers the testing, inspections, operation, and maintenance of power transformers during the entire operating lifetime, which are considered good utility practices. This instruction is summarized in the following sections.

8.1.1 Scope

Receiving, storing, installing, testing, operating, maintaining, and inspecting all power transformers with a nominal voltage of 2300 V or higher on the low-voltage winding. This instruction does not cover all details found in manufacturer's instruction book.

8.1.2 Schedule for Tests, Inspections

The schedule for all required tests and inspections is listed on *Power Transformer Acceptance and Maintenance Checklist* (see Table 8.1.8)

8.1.3 Receiving New or Rebuilt Transformers

Inspect new or rebuilt transformers promptly for manufacturing errors, or shipping damage.

- Upon delivery, prior to unloading, System Shops will inspect the transformer.
- Complete all inspections in the On Car Inspect column of *Power Transformer Acceptance and Maintenance Checklist.*
- Send copies of the completed checklist to the Regional Technical Support Engineer and the Substation Component Engineer.

8.1.4 Handling and Shipping

Follow these precautions when moving transformers:

- Lift transformers only by the means provided, and use spreaders to balance the lift when necessary.
- When lifting accessories, use lifting eyes provided for this purpose, and attach lag lines to control the load at all times.
- Always handle transformers in the upright position unless the manufacturer indicates it can be handled otherwise.
- A list of accessories and parts removed for shipping a transformer should be prepared by the disassembling crew and forwarded to the crew leader receiving the unit so that he or she can determine whether all parts have been received.
- When bushings are removed for shipment, the internal leads should

be securely tied to bushing cover plates with linen tape to prevent them from moving during shipment.

8.1.5 Storage

Bushings, radiators, or other accessories removed from transformers should be tagged by the person removing them to identify the transformer from which they are removed.

Storing Transformers

- Place transformers on sound footing in a safe or protected area.
- Seal all openings where accessories have been removed with steel shipping covers.
- Gaskets should be installed with cover plates.
- All sealed transformers should be purged and pressurized with dry nitrogen for short term storage less than 6 months.
- Refer to manufacturer's instruction book for specific instructions on storing transformers in oil or nitrogen.

Storing Bushings

- Bushings should be left in transformer if possible.
- If removed, store bushings in a clean, dry place indoors, protected from damage. Oil-filled bushings can be stored in an open area protected from weather.
- Preferred storage of oil-filled bushings is vertical, but oil reservoirs must be elevated at least 10 in. above bottom bushing stud (or other height specified in instruction book).

Storing Radiators and Other Accessories

- All openings on detached radiators must be sealed from moisture and the radiators stored so water cannot stand around the sealed openings.
- Protect all accessories from damage, moisture, and foreign materials.

8.1.6 Acceptance for Service

All transformers, old or new, moved to a substation for installation must be inspected and tested before energizing. The final acceptance for service of all transformers is the responsibility of the Regional System Maintenance Supervisor.

- Complete all sections of the Accept at Sub column of the *Power Transformer Acceptance and Maintenance Checklist* (see Table 8.1.8).

- Report unsatisfactory test results to the Technical Support Engineer and the Substation Component Engineer.

8.1.7 Safety Precautions

Follow all rules and procedures in both the Safety Rule Book and Safety Procedures for Confined Space Entry. Also follow these precautions when working on power transformers.

Oxygen Test

Since nitrogen (which cannot support life) is normally used in the gas space above the oil in transformers, an oxygen-testing device must be used to check the oxygen content prior to entry into the tank. Also, monitor the oxygen content while working inside the transformer.

Caution: *Never Enter a Transformer if the Oxygen Content is Less Than 19.5%.*

Combustible Gas Test

Before entering a transformer, test for combustible gas. The maximum acceptable level of combustible gas for safe entry is 10% of the lower explosive limit (LEL).

Carbon Monoxide Test

A carbon monoxide test should be performed before entering a transformer. Carbon monoxide (CO) should be less than 25 parts per million (ppm).

Ground to Guard Against Static Charges

Static charges can develop in the windings during electrical tests and in the transformer oil as it flows through pipes, hoses and tanks.

During oil flow into any tank, ground the tanker, filter press, piping, transformer tank and all bushings or winding leads. These grounds should be maintained for at least an hour after oil flow has ceased.

To avoid charges in the transformer windings, ground the windings before and after performing winding electrical tests. If the transformer tank is not permanently grounded, ground it temporarily during the test.

8.1.8 Internal Inspection Precautions

Internal inspections are listed in *Power Transformer Acceptance and Maintenance Checklist*. They are required for all new or rebuilt units when received, and for all units when placed in service at substations. Follow the guidelines below when making internal inspections.

- Do not open transformers during periods of fog or precipitation or when relative humidity exceeds limits outlined in the manufacturer's instruction book. If no value is specified in the instruction book, use a limit of 75%.
- Do not allow transformer to stand open any longer than is necessary because of its tendency to absorb moisture from the air.
- If the core and coils are not covered with oil, maintain a flow of dry air through the tank while it is open to help keep the insulation dry.
- For oil-filled units, lower the oil to the top of the core and coil assembly. Oil must be passed through an approved filter system. If possible, do not permit the coils and insulation to be exposed to the air.
- Check oxygen content before entry and monitor while inside the tank. Minimum acceptable level of oxygen is 19.5%.
- No one shall enter confined spaces without prior authorization and review of applicable Company safety rules and General Safety Procedures for Confined Space Entry.
- Before climbing on the transformer, remove anything that may contaminate it and cause it to fail when energized:

  Remove all items from pockets and other loose items such as watches and rings.

  Wipe shoes clean of mud, stones, and other foreign material.

  Put clean disposable booties over shoes prior to entering the transformer.

  Attach lines to tools used in tank or near open manhole so that they cannot accidentally be dropped in the transformer.
- Before replacing manhole covers, check the gaskets.
- After sealing the transformer (unless it is the open type), it should be pressurized with dry nitrogen and purged until oxygen content is less than 1%.

8.1.9 Test Procedures

Required tests and their schedules are listed for all new and existing transformers on *Power Transformer Acceptance and Maintenance Checklist.* Report any unsatisfactory results to the Regional Technical Support Engineer and the Substation Component Engineer.

Core Ground Test

The core ground test (performed on core form units only) determines the presence of unintentional ground on the transformer core by checking the core ground insulation resistance using a 1000 V megger with the factory core ground disconnected. Test values corrected to 20°C should be 25 MΩ or greater. Record results on *Power Transformer Acceptance and Maintenance Checklist*.

Transformer Turns Ratio (TTR) Test

The TTR test measures the turns ratio of the windings and detects open-circuited and short-circuited turns.

- For proper connections to transformers, refer to TTR test set instruction book.
- Perform test on all tap positions.
- Set transformer on tap to be used and retest. Results on each tap should be within 0.5% of the theoretical turns ratio.

$$\text{Theoretical ration} = \frac{\text{HV coil rating}}{\text{LV coil rating}}$$

- If satisfactory TTR test results cannot be obtained on all phases, not only the phase or phases that are unsatisfactory, note either of the following:
  High exciting current and low voltage on TTR test set (indicative of a short circuit on one of the windings)
  Normal exciting current and normal voltage but no galvanometer deflection on TTR test set (indicative of an open circuit)
- Record test results on TTR Test Results form.

Doble Test

The Doble test is performed on transformer windings and bushings by an Insulation Technician. The test is a relative measure of the condition of the insulation. The Insulation Technician will make recommendations.

Megger Test

The megger test is a measure of the insulation resistance. Remove solid connections of windings to ground. The following tests are made using a 1000 V megger:

- Measure from the high-voltage winding to the low-voltage winding and ground. (Low-voltage winding is grounded.) H-LG
- Measure from the low-voltage winding to the high-voltage winding and ground. (High-voltage winding is grounded.) L-HG
- Measure from the high- and low-voltage windings to ground. HL-G
- Megger readings are temperature dependent. Correct all the above

TABLE 8.1.1 Temperature Correction Multipliers for Insulation Resistance Measurements

Oil temperature, °C	Oil temperature, °F	Multiplier
0	32	0.25
5	41	0.36
10	50	0.50
15.6	60	0.74
20	68	1.00
25	77	1.40
30	86	1.98
35	95	2.80
40	104	3.95

readings to 20°C by multiplying the megger readings by the multipliers shown in Table 8.1.1.

- The readings for the H-LG, L-HG, and HL-G tests should be comparable to prior megger readings, with all readings corrected to 20°C (68°F).
- If previous readings are not available, the minimum acceptable resistance for each reading is per Table 8.1.2.
- Record megger readings on Megger Test Results form.

Oil Tests

Warning! To prevent air from being drawn into transformers with sealed-tank or gas-oil-seal designs, make sure the gas space above the oil is at a positive pressure before opening the drain valve.

For all transformers, perform oil dielectric and acidity tests. Refer to instruction: Insulating Oil: Using, Handling, Sampling, and Testing. Record results on *Power Transformer Acceptance and Maintenance Checklist*. For new and rebuilt transformers, send oil samples for additional tests to the Chemical Laboratory. A copy of the test results should be sent to Substation Component Engineering.

TABLE 8.1.2 Insulation Resistance Limits

Rated voltage of winding under test	Minimum acceptable resistance
2300 V to 13,200 V	800 MΩ
Above 13200 V	1600 MΩ

Oxygen Content Test

High oxygen content results in oxidation of oil, which causes sludging. The gas space in all sealed transformers is to be tested semiannually for oxygen content. Record results on *Power Transformer Acceptance and Maintenance Checklist*. Oxygen content should not exceed 5%. When necessary, the gas space should be purged with dry nitrogen until the oxygen content is 1% or less.

Excitation Test

This test is done with Doble equipment by the Insulation Technician, who will make recommendations.

Gas-in-Oil Analysis

Semiannually, samples of oil are to be taken for gas-in-oil analyses from all 69 kV and above power transformers and generating station transformers.

8.1.10 Vacuum Oil Filling

Vacuum filling is required for new or repaired transformers that are shipped without oil. Use Table 8.1.3 to determine if vacuum oil refilling is required for a transformer that has been drained or opened for inspection or maintenance.

TABLE 8.1.3 Requirements for Vacuum Oil Filling

<table>
<tr><th>If the transformer operating voltage is</th><th>And the core and coils</th><th>Then vacuum oil refilling is</th></tr>
<tr><td>115 kV and above</td><td>→</td><td rowspan="2">Required: Follow steps 1–10 below, under "Vacuum Oil Filling"</td></tr>
<tr><td rowspan="2">Below 115 kV</td><td>Have been exposed to atmosphere</td></tr>
<tr><td>Have not been exposed to atmosphere</td><td>Not required: Add oil by pumping into the top filter press connection. When oil filling is completed and prior to energizing, allow the transformer to stand. Use the chart in step 10 for the proper hold time.</td></tr>
</table>

TABLE 8.1.4 Oil Filling Requirements for Various Transformer Compartments

LTC compartment type	And the tank is being filled ("Original Fill")	And the tank is being topped off	And the tank is being filled after Maint./Repair
Selector switch	Vacuum fill	Degassed oil*	Degassed oil*
Vacuum interrupter	Vacuum fill	Degassed oil*	Vacuum fill
Diverter switch	Filtered oil	Filtered oil	Filtered oil
Arcing-type switch	Filtered oil	Filtered oil	Filtered oil
Reactor compartment	Vacuum fill	Degassed Oil*	Degassed Oil*

* Degassed Oil–Oil that has been vacuum processed prior to storage.

When oil filling an LTC compartment, use Table 8.1.4 to determine what type of oil filling is required.

When transferring oil to a power transformer, pass the oil through an approved filter system to remove moisture and sediments.

Caution: *Never Energize an Underfilled or Overfilled Power Transformer.*

Precautions for Vacuum Oil Filling

- Apply vacuum only to transformers designed for vacuum filling. (Noted on Nameplate.)
- Never stand on top of a transformer while vacuum is applied.
- Never energize a transformer while vacuum is applied.
- Some transformers require removable vacuum bracing; consult manufacturer's instructions and drawings.
- Remove pressure relief devices and sudden pressure relays and seal openings to prevent damage or misoperation due to changes in pressure when vacuum filling.

Warning! When a transformer has load tap-changing compartments, it may be necessary to apply vacuum simultaneously to equalize pressures. Consult manufacturer's instructions.

Procedure for Vacuum Oil Filling

1. Remove all oil from the transformer.
2. Test new, unprocessed oil for dielectric strength using ASTM Method D877. The oil must have a minimum breakdown voltage of 30 kV.

TABLE 8.1.5 Dew-Point Limits Based on Insulation Compartments

Insulation temp., °F	Max dew point, °F	Insulation temp., °F	Max dew point, °F	Insulation temp., °F	Max dew point, °F
0	−78	40	−44	80	−10
5	−74	45	−40	85	−6
10	−70	50	−35	90	−1
15	−66	55	−31	95	3
20	−64	60	−27	100	7
25	−58	65	−22	110	16
30	−53	70	−18	120	25
35	−48	75	−14	130	33

3. After assembly, pressurize the transformer to 2 psig by adding dry nitrogen. Check the transformer for leaks.
4. For transformers rated 115 kV and above, after waiting for a 24 h period, make a dew-point check to determine the dryness of the transformer insulation. For new transformers and transformers in warranty, refer to the manufacturer's instructions for acceptable dew-point readings. For transformers not in warranty, refer to Table 8.1.5. If it does not pass dew-point test, a hot oil dryout is required. After dryout, repeat step 4.
5. Draw a vacuum of 2 mmHg or less. Hold this vacuum for a period of time specified in the manufacturer's instruction book, if the transformer is new or in warranty. If the transformer is not in warranty, use Table 8.1.6.
6. Maintaining a vacuum of 2 mmHg or less, admit oil into the top of the transformer connection. *Once oil filling is started, it must not be interrupted. Oil degassing equipment is required for transformers rated 115 kV and above.*
7. If the transformer is a conservator type, stop filling when oil reaches a level 2 in. below the transformer cover. If the transformer

TABLE 8.1.6 Vacuum Hold Time Prior to Oil-Filling Transformer

If transformer operating voltage is	Then the vacuum hold time is
Less than 115 kV	4 h
115–345 kV	6 h
Above 345 kV	12 h

TABLE 8.1.7 Stand Time Prior to Energizing a Transformer

If transformer operating voltage is	Then allow to stand
Less than 115 kV	6 h
115–345 kV	12 h
Above 345 kV	24 h

is equipped with a nitrogen bottle, stop filling when the oil level gauge is slightly over the 25°C level. This is to compensate for the transformer expanding when vacuum is broken and for oil cooling.

8. Break the vacuum with dry nitrogen. If the transformer has a conservator with air bag, or air separation membrane, add the remaining oil to the expansion tanks in accordance with the manufacturer's recommendations.
9. Bleed the air from transformer oil pump vents. Turn on all pumps and leave them running while the oil cools.
10. Allow the transformer to stand before energizing (with oil pumps running) according to the timetable shown in Table 8.1.7. Run one half of the pumps for half the time and the other half of the pumps for the second half of the time.
11. Prior to energizing the transformer, check oil levels in all compartments. Pump oil into the top, if necessary, to raise the oil level to the 25°C mark.
12. Prior to energizing the transformer, shut off all oil pumps and place controls on automatic so that no pumps are running prior to energizing. This is important to eliminate static electrification of the oil, which could cause an internal failure. This hold time must be a minimum of 12 h.

8.1.11 Maintenance of Spare Transformers

Usable transformers stored as system spares shall receive the following minimum maintenance checks every 6 months except as noted. Failed units awaiting disposition require no periodic maintenance.

- Oil sample and dielectric test (check yearly)
- Nitrogen pressure
- Oil level
- Visual inspection, including check for oil leaks and rust
- Operation of control cabinet heaters

Table 8.1.8 Power Transformer Acceptance and Maintenance Checklist

Substation ________ Bank No. ________ Transformer No. ________
Mfr. ________ Serial No. ________ Voltage Rating ________ KVA ________

Item	On car inspect	Accept at sub	1 month	6 months	1 year
1. Megger test (enter readings below) HL-G ________ MΩ H-LG ________ MΩ L-HG ________ MΩ	√	√			
2. Core ground ________ MΩ	√	√			
3. Doble test (every 8 years)		√			
4. TTR test (all tap positions)	√	√			√
5. Oil dielectric ________ kV	√	√			√
6. Oil acidity ________ Mg KOH		√		√	√
7. Oxygen content ________%		√		√	√
8. Excitation test	√				
9. New oil test (Chemical Lab)	√	√			
10. Gas-in-oil sample				√	√
11. Oil sample	√	√			√
12. Oil temperature			√	√	√

13. Winding temperature			√	√	√
14. Load current (tong ammeter)			√	√	√
15. Voltage			√	√	√
16. Oil level	√	√	√	√	√
17. Nitrogen Pressure	√	√	√	√	√
18. Automatic gas-seal equipment		√	√	√	√
a. Pressure gauge		√	√	√	√
b. Gas cylinder content		√			√
c. Low-pressure alarm circuit		√			√
19. Fan motors—lubrication		√			√
20. Heat exchanger clear of obstruction		√			√
21. Breather openings clear		√		√	√
22. Openings, gaskets, and springs of pressure relief valves		√			
23. Pressure relief semaphore	√	√	√	√	√
24. Opening in pressure-vacuum bleeder		√			
25. Tank grounds		√	√	√	√
26. Supervisory circuits		√			
27. Check for oil leaks	√	√	√	√	√
28. Operation of sudden pressure relay, thermal overload and oil temperature indicators		√			

Table 8.1.8 Continued

Item	On car inspect	Accept at sub	1 month	6 months	1 year
29. Calibrate oil temperature indicator		√			
30. Calibrate thermal overload device (per Relay Test Dept. schedule)		√			
31. Oil pump operation and correct flow		√	√	√	√
32. Oil pump bearing wear measurement					√
33. Check for unusual noises		√	√	√	√
34. Weathertightness of control cabinet		√	√	√	√
35. Tightness of control wiring conn.		√			√
36. Check control switch operation		√	√	√	√
37. Operation and tightness of gauges, valves, and fittings		√	√	√	√
38. Check control cabinet for corrosion		√	√	√	√
39. CT connections, polarity, and ratio		√			
40. Differential relay (Test Dept. Schedule)		√			
41. Check for saturation of silica gel breather			√	√	√
42. Examine impact recorder tape	√				
43. Check for movement on RR car	√				
44. External shipping braces	√				

45. Check relative humidity before opening tank	√	√			
46. Condition of paint	√	√	√	√	√
47. Check all accessories for absence or damage	√	√			
48. Check nameplate for correctness	√	√			
Internal. Check for Oxygen Content Before Entering (Min. 19.5%)					
49. Core ground (core form units) ________ MΩ	√	√			
50. Check bushing leads for condition, length, and connections	√	√			
51. Wood-blocking and lead supports	√	√			
52. Examine all visible insulation	√	√			
53. Internal shipping braces	√	√			
54. Check for shifting of core and coils	√	√			
55. Check for loose or missing bolts, nuts, and connecting links	√	√			
56. Tap changer gear box lubrication		√			
57. Tap changer check operation	√	√			
58. Tap changer contact alignment and pressure	√	√			
59. Tap changer check operating shaft	√	√			
60. Tap changer check discoloration of current-carrying parts	√	√			
Bushings					
61. Surfaces clean		√	√	√	√

Table 8.1.8 Continued

Item	On car inspect	Accept at sub	1 month	6 months	1 year
62. Inspect for cracks	√	√	√	√	√
63. Oil level	√	√	√	√	√
64. Oil leaks	√	√	√	√	√
65. Connectors, bolts, clamps, and top terminals (every 8 years)	√	√			
66. CTs centered, insulation OK, clamps tight (every 8 years)	√	√			
67. Leads clear, not straining bushings (internal and external)	√	√	√	√	√
68. Calibrate bushing potential device (Test Dept. Schedule)		√			
69. Check for radio noise			√	√	√
Radiators					
70. Condition of paint	√	√	√	√	√
71. Check for oil leaks on fins, headers, and valves	√	√	√	√	√
72. Check bracing	√	√	√	√	√
Load tap changers—except vacuum-type LTC components in oil which are on 4-year cycle					
73. Check oil in arcing compartment—change if necessary		√			√
74. Oil sample from selector switch compartment		√			√
75. Pressure relief valve/breather clear		√	√	√	√

76. Check all contacts		√			√
77. Check motor lubrication		√			√
78. Check mechanism lubrication		√			√
79. Visual inspection of mechanism		√	√	√	√
80. Motor contactor contacts		√	√	√	√
81. Mechanism brakes		√			√
82. Gaskets—replace if necessary		√			√
83. Selsyn devices—check lubrication		√			√
84. Operation counter and position indicator		√	√	√	√
85. Clean all insulating parts		√			√
86. Check stuffing boxes for leaks		√			√
87. Operation of mechanisms		√			√
88. Tightness of all connections and nuts and bolts		√			√
89. Check for loose or damaged parts and cracked porcelain		√			√
90. Calibration of voltage regulating relays		√			
91. Contacts on voltage regulating relays and motor operating and auxiliary relays (Test Dept. Schedule)		√			

Completed by ______________ Date ______________

Work Order No. ______________ Date ______________

8.2 PREVENTATIVE MAINTENANCE VERSUS PREDICTIVE MAINTENANCE

Some of the activities mentioned in the O&M instruction above are *preventative maintenance* activities while others fall under the *predictive maintenance* category. The aim of preventative maintenance is to keep components from aging and wearing out, or to restore and replace aged or worn components before they fail. Preventative maintenance is scheduled periodically or performed on some other timetable based on past experience of the component failure modes.

The aim of predictive maintenance, on the other hand, is to detect aging or wear in components so that preventative maintenance can be performed before ultimate failure occurs. Predictive maintenance is commonly referred to as testing. The requirements for a good test are as follows:

- The test should have sensitivity; in other words, it should give an early warning of impending trouble.
- The test should have selectivity; in other words, it should not give off false positive indications of trouble and should give a clear indication of what is wrong.
- The test should be practical; in other words, it should not require exotic test equipment or an unusually high skill level to perform the test or interpret the results.
- The test should be nondestructive.

8.3 FACTORY TESTS

Every transformer that is manufactured undergoes some form of factory testing. For power transformers, these tests are quite extensive and a certain percentage of test failures do occur. Typically around 5% of the transformers produced will fail at least one of these tests. Test requirements are spelled out in a number of industry standards and specifications [2,3].

There is some overlap among industry standards with respect to testing. In recent years ANSI/IEEE Standard C57.12.90 has been generally adopted by the other standards for testing liquid-immersed distribution, power, and regulating transformers. Some of the more significant standard factory tests specified in ANSI/IEEE Standard C57.12.90 are itemized in the following sections.

8.4 RATIO TEST

This test determines the ratio (TTR) of the number of turns in the high-voltage winding to that in the low-voltage winding. The ratio test shall be made at rated or lower voltage and rated or higher frequency. In the case of three-phase transformers when each phase is independent and accessible, single-phase power should be used, although three-phase power may be used when convenient. The tolerance for the ratio test is 0.5% of the winding voltages specified on the transformer nameplate.

The accepted methods for performing the ratio test are the voltmeter method, the comparison method, and the ratio bridge. With the voltmeter method, the primary winding is excited at rated frequency and the voltage at the primary and the open-circuit voltage of the secondary winding are measured. The ratio is the primary voltage divided by the secondary voltage. The comparison method applies voltage simultaneously to the transformer under test and the open-circuit secondary voltages are measured and compared.

The ratio bridge method is the most accurate method and can easily determine the TTR to the very small tolerances required by the standard. The test apparatus is commonly referred to as a TTR Test Set. One such test set is manufactured by the Biddle Company and has proven to be especially useful as a diagnostic test in the field, so its operation will be described in detail. This test set is shown in Figure 8.1.

The clamp test leads are connected to the secondary winding of the

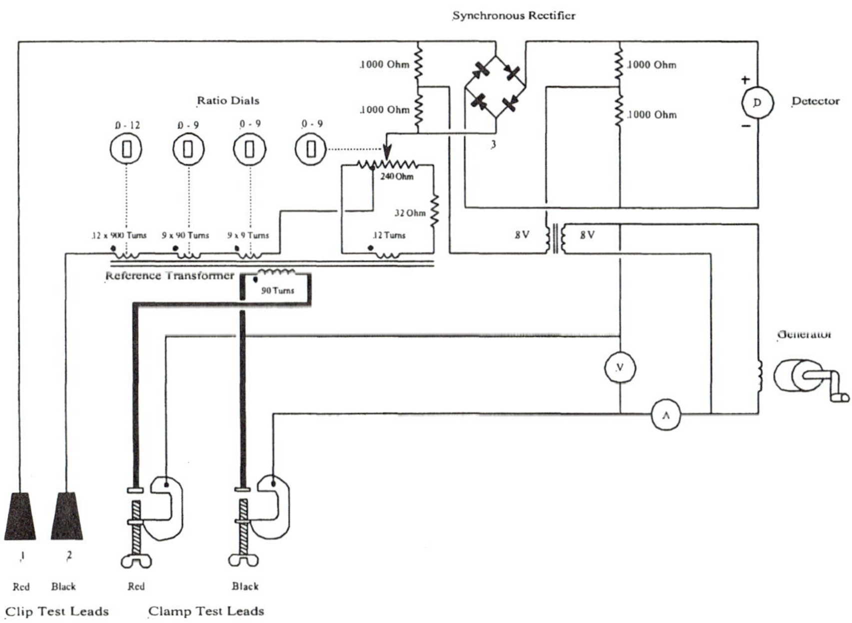

FIGURE 8.1 Circuit diagram of a TTR test set.

transformer under test and the clip leads are connected to the primary winding under test. The secondary winding of the transformer under test and the secondary of a calibrated reference transformer in the test set are both excited by the same 8 V source voltage from a hand-cranked generator. A voltmeter is used to verify that the correct voltage is being applied.

An ammeter measures the exciting current into the transformer under test.

When the voltage developed across the primary of the transformer under test (1-2) is equal to the voltage developed across the primary of the calibrated reference transformer (2-3), then the voltage across the synchronous rectifier is zero and the galvanometer detector reads zero. With more voltage developed across 1-2 than across 2-3, the galvanometer has a negative deflection. With less voltage developed across 1-2 than across 2-3, the galvanometer has a positive deflection. The ratio dials are used to adjust the ratio of the reference transformer.

A simplified equivalent circuit of the TTR test set is shown in Figure 8.2. The transformer under test is also shown. Note that the current through the detector, labeled "Det" in the figure, is zero when the voltages developed

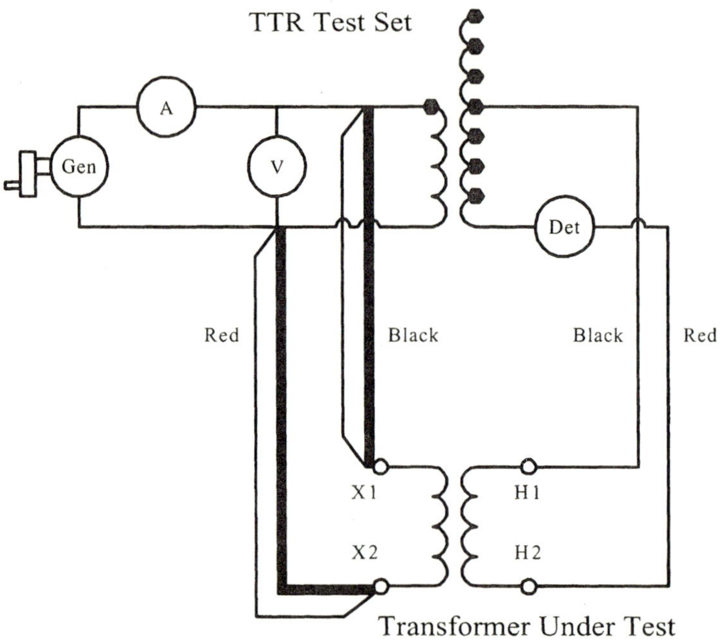

FIGURE 8.2 Simplified circuit of a TTR test set.

at the high-voltage terminals of the test-set transformer and the transformer under test are equal. This condition exists when the ratios of the test-set transformer and the transformer under test are equal.

8.5 OTHER FACTORY TESTS

The remainder of the factory tests are briefly summarized below. The details of the test set connections and formulas are beyond the scope of this book and the reader is directed to ANSI/IEEE Standard C57.12.90 for these details.

- No-Load Losses. The tests measures the no-load losses at specified excitation voltage and a specified frequency. Sine-wave voltages are used unless a different waveform is inherent in the operation of the transformer. The recommended method is the average-voltage voltmeter method, employing two parallel-connected voltmeters. One voltmeter is an average-responding but RMS calibrated voltmeter and the other voltmeter is a true RMS-responding voltmeter. The test voltage is adjusted to the specified value as read by the average-responding voltmeters. The readings of both voltmeters are used to correct the no-load losses to a sine-wave basis.
- No-Load Excitation Current. This current is measured in the winding used to excite the transformer with the other windings open-circuited. It is generally expressed in percent of the rated current of the winding. No-load excitation current is not sinusoidal and contains, as we have seen, odd harmonics (predominantly third harmonic current). The ammeter used to record the no-load excitation current is an RMS meter which reads the square root of the sum of the squares of the harmonic currents.
- Load Losses and Impedance Voltage. The transformer must be in a specific state before the load losses and impedance voltage are measured. The temperature of the insulating liquid must be stabilized and the difference between the top and bottom oil temperatures shall be less than 5°C. The winding temperatures must be measured (using a resistance method) before and after the test and the average taken as the true temperature. The difference in the winding temperature before and after the test must not exceed 5°C. The two test methods for measuring load losses and impedance voltage are the wattmeter-voltmeter-ammeter method and the impedance bridge method. These tests generally apply a reduced voltage to one set of windings with the other set of windings short-circuited. For three-winding trans-

formers, these tests are repeated for each combination of windings taken two at a time.

- Dielectric Tests. These tests consist of applied-voltage tests and induced-voltage tests. Applied-voltage tests apply a high voltage to all bushings of a winding, one winding at a time, with the other windings grounded. A 60 Hz voltage is increased gradually over 15 s and held for 40 s and reduced to zero over 5 s. Induced-voltage tests apply a high voltage across a winding with the other windings open-circuited in order to test the quality of the turn-to-turn insulation. In order to prevent core saturation at the higher excitation voltage, the frequency of the induced-voltage test is increased (typically around 120 Hz). The induced voltage is applied for 7200 cycles or 60 s, whichever is shorter.
- Switching Impulse Test. The switching impulse test applies a switching impulse wave between each high-voltage line terminal and ground. The test series consists of one reduced voltage wave (50%–70% of specified test level) followed by two full-voltage waves. Either positive or negative polarity waves, or both, may be used. A voltage oscillogram is taken for each applied wave. The test is successful if there is no sudden collapse of voltage. Successive oscillograms may differ because of the influence of core saturation.
- Lightning Impulse Test. The test sequence consists of one reduced full wave, two chopped waves, and two full waves. Tap connections are made with the minimum effective turns in the winding under tests and regulating transformers are set to the maximum buck position. Oscillograms are taken of each wave. The general technique for interpreting the results is to look for differences in the shapes of the reduced full wave and the two full waves, which indicate turn-to-turn insulation failure. Additional test criteria are found in IEEE Std. C57.98-1993. The impulse tests probably have the highest likelihood failures among all of the factory tests that are typically performed.
- Partial Discharge Test. This test detects radio-frequency (0.85–1.15 MHz) noise generated from partial discharges within voids in the insulation. An applied voltage is gradually increased until partial discharge starts to occur, which is the *inception voltage.* The voltage is then decreased until the partial discharge stops, which is the *extinction voltage*. The extinction voltage must be less than the operating voltage of the transformer; otherwise, once partial discharge starts in the field (due to some voltage transient), it would continue indefinitely and possibly cause damage or failure.

- Insulation Power Factor. Insulation power factor is the ratio of the power dissipated in the insulation in watts to the apparent power (volt-amperes) under a sinusoidal voltage. The applied 60 Hz voltage of this test is generally lower than the operating voltage of the transformer. The Doble Test Set is designed specifically to carry out this test. Portable versions are used to measure the insulation power factor of transformers in the field. This test usually must be done by a trained technician. The test results are temperature-corrected to a reference temperature of 20°C.
- Insulation Resistance. This test applies a high-voltage DC voltage to one winding at a time with the other windings grounded. The leakage current is measured and the insulation resistance is calculated using Ohm's law. A Megger Test Set is designed specifically to carry out this test, and its meter is calibrated in megohms in order that the calculation may be avoided. The Megger is a portable instrument that can easily carried around in the field.
- Noise Measurement. The noise measurement test is carried out while the transformer is energized at rated voltage with all of the cooling equipment running. Room geometry can greatly affect the measurements, so it is preferable that the transformer be inside an *anechoic chamber*; however, if such a chamber is not available, no acoustically reflecting surface may be within 3 m of the measuring microphone other than the floor or ground. The recording microphones are positioned in 1 m intervals around the perimeter of the transformer, with no fewer than four (4) microphone positions for small transformers. Sound power levels are measured over a specified band of frequencies. The sound power levels are converted into decibels (dB).
- Temperature Rise (Heat Run). The transformer is energized at rated voltage in order to generate core losses. The windings are connected to a *loading transformer* that simultaneously circulates rated currents through all of the windings in order to develop load losses. Naturally, the excitation voltage and the applied circulating currents are electrically 90° apart to minimize the KW requirements for this test. Nonetheless, a large power transformer can consume up to 1 MW of total losses and the heat run test is an expensive test to perform. Therefore, in order to reduce the total expense, heat run tests are normally performed on only one transformer on a purchase order for multiple transformers, unless the customer chooses to pay for testing additional units.

- Short-Circuit Test. The short-circuit test is generally reserved for a sample transformer to verify the design of a core and coil assembly unless the customer specifies that a short-circuit test be performed on transformers that are purchased. The customer should be cognizant of the ever-present risk of damaging the transformer during short-circuit tests. A low-voltage impulse (LVI) current waveform is applied to the transformer before and after the applications of short-circuit test. The "before" and "after" oscillograms of the LVI currents are compared for significant changes in waveshape that could indicate mechanical damage to the windings.

8.6 FIELD TESTS

In Section 8.2 we discussed the general criteria for good predictive maintenance practices. These same criteria apply to testing. There are a number of field tests that are considered good predictive maintenance practices and these should be performed periodically to spot trouble. These tests are also useful for diagnosing transformer trouble. A Megger test consists of applying a high DC voltage, usually 1000 V, to each winding with the other windings grounded and to all windings connected in parallel. The Megger readings are in megohms and these must be temperature corrected for meaningful results. The megger readings should be compared to earlier test results to detect any downward trend in resistance values. The voltage produced by a megger is high enough to cause insulation breakdown if there are gross faults, but is really not sensitive enough to detect minor problems in transformers in the higher voltage classes.

A Doble test is somewhat more sensitive than the Megger test. An AC voltage, up to 10 kV, is applied to the winding insulation and leakage current is measured. In addition to the leakage current, the power factor of the insulation is computed. A high power factor indicates lossy insulation, which can mean imminent trouble. In addition to the winding insulation, the Doble test is used to measure the power factor of bushing insulation. When testing condenser type bushings, the capacitance tap is utilized. The Doble test set is also used to measure the excitation current through the winding by applying an AC voltage across the winding. High power factor readings during this test can indicate flaws in the turn-to-turn insulation.

A TTR test can be used as a diagnostic test in the field. Always connect the TTR test set clamp leads to a secondary winding of the transformer under test. Connect the TTR test set clip leads to the primary winding that is on the same core leg as the secondary winding being tested, observing that the polarity of the red clip test lead matches the polarity of the red clamp test lead. Set the ratio dials just above zero and give the generator wheel a half turn. The galvanometer should deflect to the left, indicating the ratio dials need to be raised. A deflection to the right means that the polarity of the test leads is incorrect. This can be corrected by swapping the two clip test leads. After the correct polarity has been verified, slowly turn the generator and make the appropriate adjustments to the ratio dials in order to keep the galvanometer needle centered (zero current in the clip test leads). When the ratio dials are almost set to the right ratio, the generator can be cranked faster to get the proper voltage indication on the voltmeter (8 V).

If the voltmeter reads low voltage with the ammeter reading high current, this is usually an indication of shorted turns, either in the primary or in the secondary. A zero deflection on the galvanometer at every ratio settings indicates an open primary winding because no current can flow in the clip test leads. If the galvanometer deflection is always to the right and cannot be corrected by reversing the test leads, then this may indicate an open secondary winding and voltage cannot be generated in the primary winding.

Example 8.1

A repair crew was making a TTR test on a 500–230 kV transformer and the test leads would not stretch from the HV bushings to the LV bushings, so the crew moved the TTR Test Set closer to the HV bushings to attach the clip test leads and connected the clamp to the LV bushings by extending them using jumper wire. The voltage and current indications on the TTR Test Set were normal, but the ratio measured 2.5% lower than the theoretical current. What is wrong?

The clamp test leads are calibrated such that the resistance voltage drop from the current flow is taken into account by the measurement leads connected to the clamps. By extending the clamp leads, the crew introduced an additional resistance voltage drop that reduced the exciting voltage at the LV winding and hence the voltage induced in the primary. This gave the indication of a low TTR.

In order to perform the test correctly, the *clip* leads should have been extended. Since no current flows in the clip leads with the galvanometer reading zero when the correct ratio is dialed in, there is no voltage drop in the clip lead extensions and the reading will be accurate.

Example 8.2

A three-phase 12470Y/7200-480V station service transformer without a primary neutral bushing is to be TTR tested. How should the TTR Test Set be

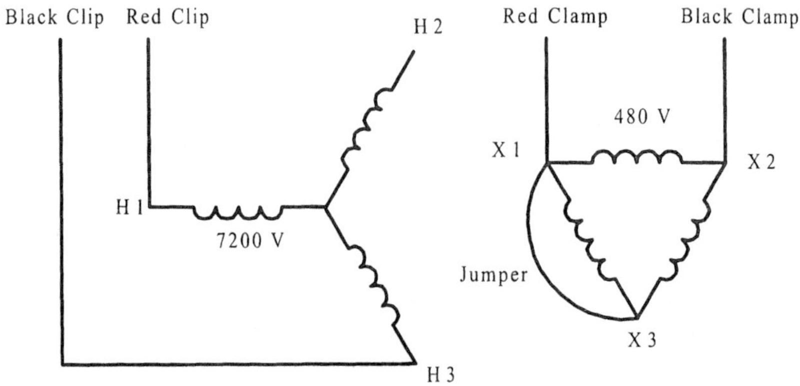

FIGURE 8.3 TTR test set connections for ratio testing an ungrounded Y-Δ transformer.

connected since the primary windings cannot be isolated? What is the TTR test reading for the nominal voltage taps?

To measure the H1-N to X1-X2 winding ratio, connect a jumper wire between X1 and X3, connect the red clip test lead to H1, the black clip test lead to H3, the red clamp test lead to X1 and the black clamp test lead to X2, as shown in Figure 8.3. The jumper wire causes the voltage across both the X1-X3 winding and the H3-N winding to be zero, so the black clip test lead is effectively connected to the primary neutral through the shorted H3-N winding.

The correct ratio is $7200/480 = 15.000 \pm 0.075$.

Example 8.3

A TTR test is to be performed on a zigzag-Δ transformer, as shown in Figure 8.4. The rated phase-to-phase voltage of the zigzag winding is 12,470 V and the phase-to-phase voltage of the Δ secondary is 2,400 V. Show *all* TTR test set connections required to test the primary to secondary ratio on the *A* phase leg of the primary windings. (*Hint*: Two separate sets of connections are required to complete the test. The use of jumper wires is permitted.) What are the expected readings on the TTR test set?

The jumpers effectively short out each half of the phase-to-neutral winding. For a zigzag connection, each remaining half of the winding is rated 0.5774 times the phase-to-neutral voltage:

$$0.5774 \times 7200 \text{ V} = 4154.4 \text{ V}$$

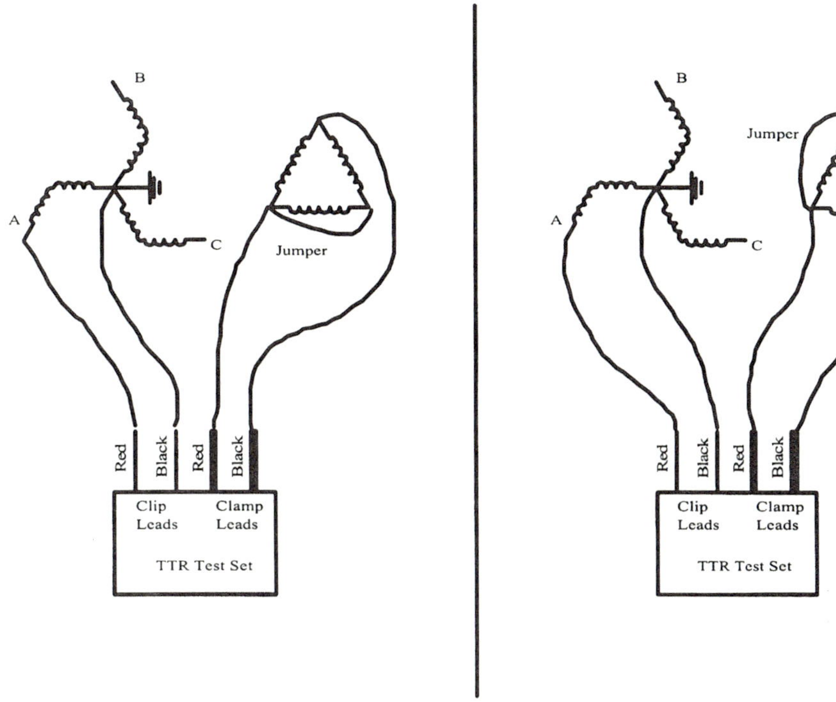

FIGURE 8.4 TTR test set connections for ratio testing a zigzag-Δ transformer.

The secondary windings are rated 2400 V. The measured turns ratio is 4157/2400 = 1.732:1.

8.7 GAS-IN-OIL ANALYSIS

Various research organizations, such as Westinghouse Electric Corporation, Analytical Associates, Inc., that did extensive research in the 1970s quickly led to the widespread use of dissolved gas-in-oil analysis as a predictive maintenance tool [4]. There is also an extensive bibliography on this subject found in IEEE Std. C57.104–1991 [5].

The basic theory is straightforward: Transformer dielectric fluids are refined from petroleum and are very complex mixtures containing aromatic, naphthenic, and paraffinic hydrocarbons. At high temperatures, some of these molecules break down into hydrogen plus small hydrocarbon molecules such as, methane, ethane, ethylene, acetylene, propane, and propylene. This process is known as *cracking*.

The kraft paper materials that are used to insulate transformer windings are made up of cellulose. At high temperatures, cellulose oxidizes to form carbon dioxide (CO_2), carbon monoxide (CO) and water (H_2O). High concentrations of CO_2 and or CO are indications of overheated windings.

All of the breakdown products are gases that dissolve readily in transformer oil in different concentrations, depending on the specific gas and the temperatures that produce them. By taking samples of transformer insulating oil, extracting the dissolved gases and doing a quantitative analysis of the various gases in the samples through *gas chromatography*, it is possible to infer the temperatures at the sites where these gases were produced.

At temperatures below 150°C, transformer oil starts breaking down into methane (CH_4) and ethane (C_2H_6). At temperatures above 150°C, ethylene (C_2H_4) begins to be produced in large quantities while the concentration of ethane decreases. At around 600°C, the ethylene production peaks while the concentration of methane continues to increase. Acetylene (C_2H_2) production starts at around 600°C and methane concentration peaks at 1000°C. Hydrogen (H_2) production is not significant below 700°C and continues to increase along with acetylene at temperatures above 1400°C.

Therefore, the relative concentrations of the key gases change over a wide range of temperature. This is basis for the application of dissolved gas-in-oil analysis for predictive and diagnostic use. An approximate formula uses the ratio of C_2H_4/C_2H_6 to derive the temperature of oil decomposition between 300°C and 800°C:

$$T(°C) = 100 \times \frac{C_2H_4}{C_2H_6} + 150$$

The so-called *Rogers ratio method* takes the ratios of several key gases into account to develop a code that is supposed to give an indication of what is causing the evolution of gas. The codes for the four-ratio method are given in Table 8.2. A fairly detailed diagnosis of transformer trouble can be derived from various combinations of codes, shown in Table 8.3.

The diagnoses shown above were derived from empirical observation. The problem with the four-ratio Rogers code is that a code generated from the gas concentrations will often not match any of the "known" diagnoses. So like a rare disease with strange symptoms, many cases of transformer trouble cannot be diagnosed at all using this method. Another method, called the three-ratio method, sometimes works when the four-ratio method does not.

TABLE 8.2 Four-Ratio Code

Ratio	Ratio range	Code
CH_4/H_2	≤ 0.1	5
	0.1–1	0
	1.0–3.0	1
	≥ 3	2
C_2H_6/CH_4	<1.0	0
	≥ 1.0	1
C_2H_4/C_2H_6	<1.0	0
	1.0–3.0	1
	>3.0	2
C_2H_2/C_2H_4	<0.5	0
	0.5–3.0	1
	>3.0	2

In the three-ratio method, the values of A, B, and C are given in Table 8.4 with the corresponding diagnoses for the various combinations given in Table 8.5. Not only are the ratios of the key gases important, but the total quantity of dissolved gas and the rate of increase are also important factors in making a diagnosis. One of the criteria for making a judgment call is the *total combustible* gas concentration. The combustible gases include H_2, CH_4,

TABLE 8.3 Diagnosis of Transformer Faults Based on the Four-Ratio Code

$\frac{CH_4}{H_2}$	$\frac{C_2H_6}{CH_4}$	$\frac{C_2H_4}{C_2H_6}$	$\frac{C_2H_2}{C_2H_4}$	Diagnosis
0	0	0	0	Normal deterioration
5	0	0	0	Partial discharge (corona)
1,2	0	0	0	Slight overheating below 150°C
1,2	1	0	0	Overheating 150°C–200°C
0	1	0	0	Overheating 200°C–300°C
0	0	1	0	General conductor overheating
1	0	1	0	Winding circulating currents
1,2	0	2	0	Core and tank circulating currents, overheated joints
0	0	0	1	Flashover without power follow through
0	0	1,2	1,2	Arc with power follow through
0	0	2	2	Continuous sparking to floating potential
5	0	0	1,2	Partial discharge with tracking

TABLE 8.4 Three-Ratio Code

	A	B	C
Ratio range	$\frac{C_2H_2}{C_2H_4}$	$\frac{CH_4}{H_2}$	$\frac{C_2H_4}{C_2H_6}$
<0.1	0	1	0
0.1–1	1	0	0
1–3	1	2	1
>3	2	2	2

C_2H_4, C_2H_6, C_2H_2, which are produced by oil decomposition, and CO, which is produced by cellulose decomposition. Each utility has a different philosophy and a different threshold for concern. Table 8.6 gives one set of guidelines based on good utility practice that is useful for determining the overall health of a power transformer based on the total concentration of combustible gases.

It is generally accepted that if the rate of combustible gas generation exceeds 100 ppm per day on a continuing basis, or if the presence of C_2H_2 exceeds 20 ppm, then consideration should be given to taking the transformer out of service to perform additional tests and inspection. IEEE Std. C57.104-1991 Table 3 also provides a set of actions based on the total dissolved combustible gas (TDCG) concentrations as well as the daily rate of TDCG production. According to the IEEE Guide, a rate of 30 ppm per day is the threshold for considering removing the transformer from service.

Oil samples are taken from the bottom drain valve in a sealed syringe to prevent the dissolved gases from escaping. The samples are sent to a chemical

TABLE 8.5 Diagnosis of Transformer Faults Based on the Three-Ratio Code

A	B	C	Diagnosis
0	0	0	No fault
0	1	0	Partial discharge of low energy
1	1	0	Partial discharge of high energy
1, 2	0	1, 2	Discharge of low energy, arcing
1	0	2	Discharge of high energy, arcing
0	0	1	Thermal fault 150°C, conductor overheating
0	2	0	Thermal fault 150°C–300°C, mild oil overheating
0	2	1	Thermal fault 300°C–700°C, moderate oil overheating
0	2	2	Thermal fault 700°C, severe oil overheating

TABLE 8.6 Recommended Actions Based on Total Combustible Dissolved Gas in Oil

0 to 500 ppm	Represents normal aging. Analyze again in six months.
501–1200 ppm	Indicates decomposition may be in excess of normal aging. Analyze again in three months.
1201–2500 ppm	Indicates more than normal decomposition. Analyze again in one month.
More than 2501 ppm	Make weekly analysis to determine rate of generation. After establishment of rate of generation, contact manufacturer concerning future operation.

laboratory where the dissolved gases are extracted from the sample under vacuum and analyzed using a *gas chromatograph*. The results are reported as ppm dissolved in oil.

Example 8.4

(*This example is based on an actual event.*)

A generator step-up transformer had been repaired in a transformer shop. The first gas-in-oil sample was taken approximately 1 month after the transformer returned to service and the sample showed over 1000 ppm of total combustible gas with over 10 ppm of acetylene. Neither the four-ratio nor the three-ratio code matched any of the diagnoses, but the key gases indicated a bare-metal hot spot (no cellulose involvement). After 1 week a second oil sample showed a rate of increase of combustible greater than 200 ppm per day. A decision was made to shut down the plant and take the transformer out of service. An inspection quickly revealed the cause: The neutral ends of the high-voltage Y-connected windings were attached to a neutral wire that was badly overheated and nearly burned completely through. An investigation revealed that the neutral conductor was grossly undersized. Apparently, the designer believed that the "neutral" current was nearly zero because the phases are balanced, without realizing the conductor *between* the connections to the phases carries the full phase currents!

8.8 WATER-IN-OIL ANALYSIS

There is an old expression, "Oil and water do not mix." Thus, oil is not usually thought of as having a great affinity for water, and in fact it doesn't. However, the kraft paper insulation found in most power transformers has a tremendous affinity for water. In fact, cellulose is often used as a drying agent or desiccant. If there is moisture present in the transformer, it will usually wind up in the kraft paper insulation. Moisture not only weakens the insulating properties of the kraft paper, it also accelerates the rate of aging. Therefore, in order to prolong the life of a transformer, moisture must be monitored.

Since samples of the insulation cannot be taken while the transformer is in service, water-in-oil analysis is used to monitor the moisture content of the kraft paper as a surrogate. There is a known equilibrium between moisture concentrations in the kraft paper versus the moisture concentrations in the oil based on the temperature of the paper and oil. The equilibrium is expressed by the so-called Piper chart, shown in Figure 8.5 [6].

Notice that as the temperature increases, water is driven from the paper into the oil. At elevated temperatures the oil is able to dissolve more water than at lower temperatures. The relationship can be expressed by the following equation.

$$T = 31.52 - 26.605 \text{ Ln pct} + 17.524 \text{ Ln ppm} \tag{8.8.1}$$

where

T = temperature (°C)
pct = % water in paper
ppm = ppm water in oil

When doing a water-in-oil analysis, a syringe sample of oil is taken from the drain valve. *Care must be exercised so that the oil is not exposed to the atmosphere*. (Any exposure to the atmosphere will cause the oil to

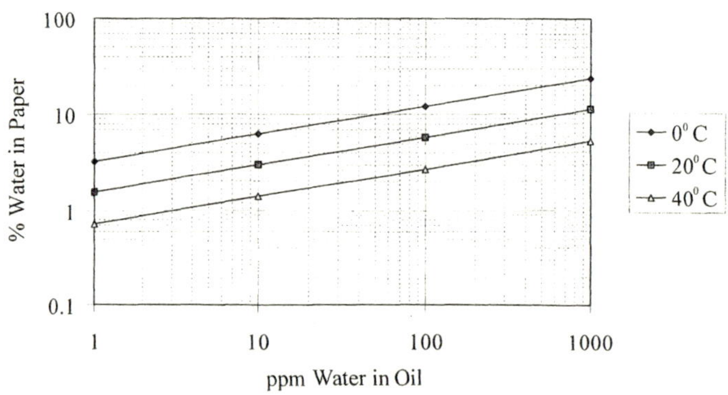

FIGURE 8.5 Piper equilibrium chart. (From Ref. 6.)

quickly reach equilibrium with the air. Since ambient air usually contains quite a bit of moisture, this will generally immediately saturate the oil with water and produce a meaningless analysis.) The oil temperature is recorded at the time the sample is taken and the sample is then sent to a chemical laboratory to analyze the ppm water in oil. From the ppm in the oil sample and the temperature of the oil, the Piper chart or Eq. (8.8.1) can be used to get an approximate indication of the percent moisture in the kraft paper.

Note that the temperature of the oil/paper interface has a significant effect on the equilibrium moisture concentration, but the temperatures of the oil and the paper vary depending on location. We would then expect the equilibrium moisture concentration to vary as well, which it does. Generally, the insulation near the hottest spot will have less percent moisture than insulation exposed to cooler oil at the bottom of the transformer. An "average" value of the percent moisture concentration could be calculated from an "average" temperature; however, this may result in a misleading assessment of the transformer's state because of the wide variation in moisture concentrations. A conservative assessment would base the percent moisture on the oil temperature at the bottom of the tank. According to the Transformer Maintenance Institute, 2% is the absolute upper limit for acceptability for percent moisture in kraft paper. Generally, if the percent moisture is less than 1%, the transformer is considered "dry."

There is also an equilibrium equation between vapor pressure of water in air (humidity) and % water in paper.

$$T = 40.17 - 22.285 \text{ Ln pct} + 14.056 \text{ Ln vap} \tag{8.8.2}$$

where vap = vapor pressure, mmHg.

Since the dew point of air is related to the vapor pressure, a dew-point measurement of the space inside a transformer before oil filling is a very good indication of the amount of water locked in the paper. This will determine whether oil filling should proceed or further drying is necessary.

Example 8.5

An oil sample is taken in an open jar from a transformer with a top oil temperature of 70°C at an ambient temperature of 25°C and 80% relative humidity. The oil sample is analyzed and it is found that there is 32.5 ppm water in oil. Using a piper chart, it is determined that there was 2.33% water in the paper, which is unacceptable. In reality, however, it was the air that produced the 32.5 ppm water in oil because this is the equilibrium ppm at 25°C and 80% relative humidity.

8.9 DRYING TRANSFORMERS

If it has been determined that a transformer has an excessive moisture content, either through water-in-oil analysis or dew-point analysis, the transformer should be dried out. Otherwise, the presence of moisture in the kraft paper will definitely shorten the life due to chemical decomposition through oxidation of the cellulose.

Drying out a transformer can be accomplished by untanking the core and coil assembly and drying it in an oven, but this would be both expensive and inconvenient. Fortunately, there are methods that can be applied in the field that are very effective in drying out a transformer.

A simple hot oil flush is one method, although it is fairly slow. Oil is pumped through a filter press, which contains paper filter elements, and then heated and sprayed onto the core and coils. The hot oil has a greater affinity for moisture than cold oil and will pick up some moisture from the core and coil insulation. Running the oil through a filter press allows the filter elements to remove some of the moisture in the oil. Cooling the oil before it goes through the filter press will enhance the removal of moisture from the oil.

A better method is to spray hot oil into the transformer under vacuum. The combination of the heat from the hot oil and the vacuum will cause the moisture in the insulation to "flash off" and dry the transformer. A *cold trap* can enhance the action of a mechanical vacuum pump. Once the mechanical pump removes all the air from the transformer tank, the only gas left in this space is H_2O. A cold trap using a dry ice and acetone jacket will instantly freeze any water vapor with the low-temperature ice having a very low vapor pressure. The cold trap thus acts like an auxiliary vacuum pump. The ice provides a direct measurement of the amount of moisture removed from the cellulose.

8.10 OIL DIELECTRIC TEST

Another predictive maintenance activity is periodic performance of oil dielectric testing. An oil sample is taken and placed in a standard test cup with flat electrodes. The electrodes are separated by a calibrated distance (ASTM D877 Method). An increasing 60 Hz AC voltage is applied to the electrodes until the oil "breaks down," tripping out the test set. The test cup is flushed out and the test is repeated a number times with oil from the same sample. An average value of the breakdown voltage is determined from these tests. Generally, the oil is satisfactory if the average breakdown voltage is 30 kV or higher.

Oil can lose insulating properties from excessive moisture, acids, sludge, and other contaminants. A dielectric test can quickly determine whether or not the oil is suitable for use as an insulating fluid in a transformer without performing extensive chemical tests on the oil.

8.11 RELIABILITY CALCULATIONS

Equipment failures do occur even with the best equipment designs available and the using the best utility practices. In order to operate a power system reliably, transformer failures must be anticipated and planned for by including a certain number redundant or spare transformers to the in-service transformers. Allocating a single spare for each transformer size, based on the deterministic approach of a single failure, is clearly inadequate because of the long lead times associated with procuring a replacement transformer. Multiple failures within the periods defined by these lead times are a distinct possibility, so that more than one spare for each transformer size may be necessary. The question is, "How many?" This question is answered in the remaining sections of this chapter.

8.12 THE CHI-SQUARED DISTRIBUTION

The first order of business is to determine the *failure rate* λ, expressed as failures per unit-year in service. This can be accomplished by observing a population of transformers over time. Since utility companies have rather large transformer populations that have been in service for many years, this is simply a matter of keeping good records of when each transformer goes into service, when it is removed from service, and when it fails. The simplest way to calculate λ is to divide the total number of failures in the population f by the total number of transformer years in-service T.

$$\lambda = \frac{f}{T} \tag{8.12.1}$$

Since the number of failures for a given population is rather small (we hope), there is always some uncertainty as to whether the failure rate calculated by Eq. (8.12.1) is truly representative of the population. A better estimate of λ is obtained from the chi-squared probability distribution, symbolized by $\chi^2_{n,1-\alpha}$. The parameter n is the number of *degrees of freedom*, and the parameter α is some *confidence limit*. The number of degrees of freedom is equal to 2 times the number of failures that were observed in the population. The confidence limit is equal to the probability that the true value of λ will be bounded by the calculated value.

$$\lambda = \frac{\chi^2_{2f,1-\alpha}}{2T} \tag{8.12.2}$$

Values of the chi-squared distribution for various degrees of freedom and probability values are given in Table 8.7. The probability values are equal to $1 - \alpha$.

TABLE 8.7 Chi-Squared Distribution Values

Probability/ Degrees of Freedom	0.01	0.05	0.1	0.2	0.4	0.5	0.6	0.8	0.9	0.95	0.99
1	6.6349	3.8415	2.7055	1.6424	0.7083	0.4549	0.2750	0.0642	0.0158	0.0039	0.0002
2	9.2104	5.9915	4.6052	3.2189	1.8326	1.3863	1.0217	0.4463	0.2107	0.1026	0.0201
3	11.3449	7.8147	6.2514	4.6416	2.9462	2.3660	1.8692	1.0052	0.5844	0.3518	0.1148
4	13.2767	9.4877	7.7794	5.9886	4.0446	3.3567	2.7528	1.6488	1.0636	0.7107	0.2971
5	15.0863	11.0705	9.2363	7.2893	5.1319	4.3515	3.6555	2.3425	1.6103	1.1455	0.5543
6	16.8119	12.5916	10.6446	8.5581	6.2108	5.3481	4.5702	3.0701	2.2041	1.6354	0.8721
7	18.4753	14.0671	12.0170	9.8032	7.2832	6.3458	5.4932	3.8223	2.8331	2.1673	1.2390
8	20.0902	15.5073	13.3616	11.0301	8.3505	7.3441	6.4226	4.5936	3.4895	2.7326	1.6465
9	21.6660	16.9190	14.6837	12.2421	9.4136	8.3428	7.3570	5.3801	4.1682	3.3251	2.0879
10	23.2093	18.3070	15.9872	13.4420	10.4732	9.3418	8.2955	6.1791	4.8652	3.9403	2.5582
11	24.7250	19.6752	17.2750	14.6314	11.5298	10.3410	9.2373	6.9887	5.5778	4.5748	3.0535
12	26.2170	21.0261	18.5493	15.8120	12.5838	11.3403	10.1820	7.8073	6.3038	5.2260	3.5706
13	27.6882	22.3620	19.8119	16.9848	13.6356	12.3398	11.1291	8.6339	7.0415	5.8919	4.1069
14	29.1412	23.6848	21.0641	18.1508	14.6853	13.3393	12.0785	9.4673	7.7895	6.5706	4.6604
15	30.5780	24.9958	22.3071	19.3107	15.7332	14.3389	13.0298	10.3070	8.5468	7.2609	5.2294
16	31.9999	26.2962	23.5418	20.4651	16.7795	15.3385	13.9827	11.1521	9.3122	7.9616	5.8122
17	33.4087	27.5871	24.7690	21.6146	17.8244	16.3382	14.9373	12.0023	10.0852	8.6718	6.4077
18	34.8052	28.8693	25.9894	22.7595	18.8679	17.3379	15.8932	12.8570	10.8649	9.3904	7.0149
19	36.1908	30.1435	27.2036	23.9004	19.9102	18.3376	16.8504	13.7158	11.6509	10.1170	7.6327
20	37.5663	31.4104	28.4120	25.0375	20.9514	19.3374	17.8088	14.5784	12.4426	10.8508	8.2604
21	38.9322	32.6706	29.6151	26.1711	21.9915	20.3372	18.7683	15.4446	13.2396	11.5913	8.8972
22	40.2894	33.9245	30.8133	27.3015	23.0307	21.3370	19.7288	16.3140	14.0415	12.3380	9.5425
23	41.6383	35.1725	32.0069	28.4288	24.0689	22.3369	20.6902	17.1865	14.8480	13.0905	10.1957
24	42.9798	36.4150	33.1962	29.5533	25.1064	23.3367	21.6525	18.0618	15.6587	13.8484	10.8563
25	44.3140	37.6525	34.3816	30.6752	26.1430	24.3366	22.6156	18.9397	16.4734	14.6114	11.5240
26	45.6416	38.8851	35.5632	31.7946	27.1789	25.3365	23.5794	19.8202	17.2919	15.3792	12.1982
27	46.9628	40.1133	36.7412	32.9117	28.2141	26.3363	24.5440	20.7030	18.1139	16.1514	12.8785
28	48.2782	41.3372	37.9159	34.0266	29.2486	27.3362	25.5092	21.5880	18.9392	16.9279	13.5647
29	49.5878	42.5569	39.0875	35.1394	30.2825	28.3361	26.4751	22.4751	19.7677	17.7084	14.2564
30	50.8922	43.7730	40.2560	36.2502	31.3159	29.3360	27.4416	23.3641	20.5992	18.4927	14.9535

Example 8.6

A population of transformers had three observed failures over a total of 750 transformer-years in service. What is the failure rate within a 95% confidence limit?

$$\alpha = 0.95$$
$$2T = 2 \times 750 \text{ years} = 1500 \text{ years}$$
$$2f = 2 \times 3 \text{ failures} = 6$$

From Table 8.7,

$$\chi^2_{2f,1-\alpha} = \chi^2_{6,0.05} = 12.592$$

From Eq. (8.12.2),

$$\lambda = \frac{12.592}{1500} = 0.0084 \text{ failures per transformer year}$$

Based on a 95% confidence limit, the true value of λ will not exceed 0.0084 failures per transformer year.

8.13 THE POISSON DISTRIBUTION

The probability that a certain number of failures n will occur within a certain time period is given by the Poisson distribution if the failures occur randomly and the failed units are replaced. If every failed transformer is replaced by a new unit, it can be shown that the population will eventually reach a static age distribution as the replacement process continues over a long time and once a static age distribution is reached, the failure rate of the population remains constant [7].

The Poisson distribution is defined by a single parameter μ, which is the mean value of all possible values of $n \geq 0$.

$$P(n) = \frac{e^{-\mu}\mu^n}{n!} \tag{8.13.1}$$

For a population of transformers undergoing failures and replacements, μ is equal to the mean number of failures occurring within some specific time interval and $P(n)$ is the probability that exactly n failures actually do occur within that time interval.

Let M equal the number of in-service transformers and let mttr equal the mean time to replace a failed transformer. For a population of transformers having a static age distribution and a failure rate λ, the mean number of transformers that fail during the interval mttr is given by

$$\mu = M \times \lambda \times \text{mttr} \tag{8.13.2}$$

The number of transformers that fail within the period mttr is equal to the number of transformers that exist in the under-repair status. Since the probability of exactly n transformers failing within the period mttr is determined by the Poisson distribution, the probability of exactly n units existing in the under-repair status at any given time is equal to $P_r(n)$, where

$$P_r(n) = \frac{e^{-\mu_r}\mu_r^n}{n!} \tag{8.13.3}$$

where μ_r = the mean number of transformers under repair. Note that since the mean number of transformers under repair is equal to the number of transformers that fail during the period mttr, it follows that μ_r is given by Eq. (8.13.2).

If there are N spare transformers assigned to the in-service population, the probability that all N units are depleted, P_u, is equal to the probability that $n \geq N$.

$$P_u = P_r(n \geq N) = 1 - P_r(n < N) \tag{8.13.4}$$

$$P_u = 1 - \sum_{n=0}^{N-1} \frac{e^{-\mu_r}\mu_r^n}{n!} \tag{8.13.5}$$

Let MTBF_u equal the mean time between transformer failures in the population when all of the spares have been depleted. This is equal to the mean time between in-service transformer unavailabilities.

$$\text{MTBF}_u = \frac{1}{\text{M} \times \lambda \times P_u} \tag{8.13.5}$$

Example 8.7

There are 100 transformers of a certain type in service, which have a failure rate of 0.015 failures per transformer year in service. The mean time to repair a failed transformer or procure a new transformer is 6 months (0.5 year). How many spare transformers need to be allocated in order that the mean time between transformer unavailabilities is greater than 50 years?

From Eq. (8.13.2),

$$\mu_r = 100 \times 0.015 \times 0.5 = 0.75 \text{ units under repair}$$

From Eq. (8.13.5),

$$\mathrm{MTBF}_u = \frac{1}{M \times \lambda \times P_u} > 50 \text{ years}$$

Solving for P_u,

$$P_u < \frac{1}{M \times \lambda \times 50} = \frac{1}{100 \times 0.015 \times 50} = 0.0133$$

Using Eq. (8.13.5) to calculate P_u as a function of N with $\mu_r = 0.75$, the following tabulation results.

$N = 0$: $P_u = 1.000 > 0.0133$

$N = 1$: $P_u = 0.528 > 0.0133$

$N = 2$: $P_u = 0.173 > 0.0133$

$N = 3$: $P_u = 0.041 > 0.0133$

$N = 4$: $P_u = 0.007 < 0.0133$

$N = 4$ spare transformers is the minimum number that satisfies the condition $P_u < 0.0133$ and $\mathrm{MTBF}_u > 50$ years.

8.14 STATISTICAL ECONOMICS

We are often faced with a decision to purchase a new transformer among different brands having different initial costs and different failure rates. Occasionally, we must decide whether to purchase a transformer whether to purchase an extended warranty for a new transformer or take our chances of having a failure. An economic decision can be made on the basis of failure statistics, hence the term *statistical economics*.

Suppose a particular type of transformer has a known failure rate λ. The probability that a unit fails between a time t and $t + dt$ in the future is given by the following equation:

$$dP(t) = e^{-\lambda t}\lambda\, dt \tag{8.14.1}$$

Let $C(0)$ equal today's cost to repair or replace a failed transformer, and let $C(t)$ equal the cost to repair or replace a failed transformer at t years in the future.

$$C(t) = C(0)e^{xt} \tag{8.14.2}$$

where x = inflation rate per year. The present-worth value of $C(t)$, based on an inflation rate i is given by

$$C_P(t) = C(t)e^{-it} = C(0)e^{(x-i)}t \tag{8.14.3}$$

The incremental expected present-worth cost of repair for a failure that occurs between t and $t + dt$ is equal to $C_P(t)$ times the probability that this failure actually occurs.

$$dC_P(t) = dP(t) \times C_P(t) = C(0)\lambda e^{-\lambda t} e^{-(i-x)t} \tag{8.14.4}$$

$$dC_P(t) = C(0)\lambda e^{-(\lambda+i-x)t} \tag{8.14.5}$$

Normally, the transformer manufacturer provides a standard warranty for W years at no extra charge. Suppose L is the nominal life of the transformer. The expected present-worth cost to replace this transformer over the *out-of-warranty life* of the transformer is found by integrating $dC_P(t)$ from $t = W$ to $t = L$.

$$C_P = C(0) \int_W^L \lambda e^{-(\lambda+i-x)t}\, dt \tag{8.14.6}$$

$$C_P = \frac{C(0)\lambda}{\lambda + i - x}\left(e^{-W(\lambda+i-x)} - e^{-L(\lambda+i-x)}\right) \tag{8.14.7}$$

Manufacturers often offer extended warranties to cover transformer failures that occur beyond the term of the standard warranty. Purchasing an extended warranty is justified if the price paid is less than the expected cost of a transformer failure that occurs between the time the standard warranty expires and the time the extended warranty expires. If the failure occurs before or after this period, the extended warranty is worthless. The price breakpoint can be found from Eq. (8.14.7) by setting L = the extended warranty expiration.

Example 8.8

Two brands of transformers are evaluated on an economic basis. Brand X has a purchase price of \$50,000 with a 2-year warranty. Based on past performance, brand X is expected to have a failure rate of 0.01 per transformer year in-service. Brand Y has a purchase price of \$48,000 with a 1 year warranty. Based on past performance, brand Y is expected to have a failure rate of 0.015 per transformer year in-service. Clearly, brand X is more reliable and has a better warranty, but is purchasing brand X justified? Assume an inflation rate of 0.04 per year and an interest rate of 0.09 per year, and use a 20-year life.

For brand X, $\lambda = 0.01$, $C(0) = \$50{,}000$.

$$\lambda + i - x = 0.01 + 0.09 - 0.04 = 0.06 \text{ per year}$$

Using Eq. (8.14.7) for $W = 2$ years and $L = 20$ years,

$$C_P = \frac{\$50{,}000 \times 0.01}{0.06} \times (e^{-2\times(0.06)} - e^{-20\times(0.06)}) = \$4881$$

For brand Y, $\lambda = 0.015$, $C(0) = \$48{,}000$.

$$\lambda + i - x = 0.015 + 0.09 - 0.04 = 0.065 \text{ per year}$$

Using Eq. (8.14.7) for $W = 1$ year and $L = 20$ years,

$$C_P = \frac{\$48{,}000 \times 0.01}{0.06} \times (e^{-1\times(0.06)} - e^{-20\times(0.06)}) = \$7361$$

The \$2000 difference in the initial purchase price is more than offset by the difference in expected out-of-warranty replacement costs reflected in the values of C_P for brands X and Y calculated above. Therefore, brand Y is the preferred choice.

Example 8.9

A new transformer is purchased for \$100,000. The manufacturer gives a standard 1 year warranty free of charge but also offers an extended warranty through 4 years for an added price of \$4000. Based on previous experience, the failure rate of this unit is expected to be 0.7% during the first 10 years in service. Assuming an inflation rate of 5% per year and an interest rate of 10% per year, should the extended warranty be purchased from the manufacturer?

$$\lambda + i - x = 0.007 + 0.10 - 0.05 = 0.057 \text{ per year}$$

Using Eq. (8.14.7) with $W = 1$ and $L = 4$,

$$C_P = \frac{\$100{,}000 \times 0.007}{0.057} (e^{-1\times(0.057)} - e^{-4\times(0.057)}) = \$1823$$

The expected present worth of the extended warranty is \$1823, whereas its price is \$4000; therefore, the extended warranty is overpriced and it should not be purchased.

REFERENCES

1. Power Transformers: Acceptance, O&M, Tests, Inspections. Distribution Department Instructions DDI S-110. Pennsylvania Power & Light Company, Allentown, PA, 1997.

2. IEEE Std. C57.12.90-1999. IEEE Standard Test Code for Liquid-Immersed Distribution, Power, and Regulating Transformers. Institute of Electrical and Electronics Engineers, Inc., New York, 1999.
3. IEEE Std. C57.12.00-2000. IEEE Standard General Requirements for Liquid-Immersed Distribution, Power, and Regulating Transformers. Institute of Electrical and Electronics Engineers, Inc., New York, 2000, pp. 44–48.
4. F. Jakob and T. J. Haupert. Dissolved Gas Analysis, DGA. A Preventative Maintenance Tool. Training handout by Analytical Associates, Inc.
5. IEEE Std. C57.104-1991. IEEE Guide for the Interpretation of Gases Generated in Oil-Immersed Transformers. Institute of Electrical and Electronics Engineers, Inc., New York, 1991, pp. 9–27.)
6. S. D. Myers, J. J. Kelley, and R. H. Parrish. A Guide to Transformer Maintenance. S. D. Myers, Inc., Akron, OH, 1988, p. 257.
7. J. J. Winders, Jr. System Reliability Optimization Through Statistical Economics. Sixteenth Inter-RAM Conference for the Electric Power Industry Proceedings. Monterey, CA, 1989, pp. 100–104.

Index

Acetylene, 262, 264
Air gap, 8–10
  elimination of, 16, 79–80
  force across, 10
  preventative autotransformer, 206
Ampere turns (*see* Magnetomotive force)
Ancillary equipment, 189–220
  bushings, 189–192
    bulk type, 191
    draw lead, 190–191
    interchangeable, 192
    storage, 237
  cooling equipment, 196–201
    forced air, 198
    forced oil, 198–199
    loss of, 218–220
    maintenance of, 200
    radiator, 199–200, 237
  load tap changer, 204–209

[Ancillary equipment]
    arcing contact, 206–208
    automatic controls, 210–212
    maintenance of, 208
    moving contact, 205–208
    preventative autotransformer, 205–208
    resistance type, 208–209
    transfer switch, 207–208
  oil level gauge, 201
  oil preservation, 192–196
    conservator tank, 193
    desiccant, 194
    expandable bladder, 194
    gas oil seal, 194
    Inertaire, 195–196
    sealed-tank design, 193
  pressure gauge, 201–202
  tap changer for deenergized operation, 202–204

[Ancillary equipment]
rotary drum, 202–203
temperature gauge, 201
Anechoic chamber, 257
Apparent power, 53
Asymmetrical peak current, 149, 154, 183–184
Attenuation factor, 179
Autotransformer connection, 117–133
advantages of, 132–133
boosting, 117–121
bucking, 118, 120–121
capacity multiplication, 119–122
common winding, 118
design, 131–132
disadvantages of, 133
impedance, 122–128
impulse voltages, 131
limitations, 128–129, 133
series winding, 118, 132
short circuit, 129–130, 132, 147–148

B-H curve, 10–14, 15
Banking single-phase transformers, 48–50, 93–98
Biddle TTR test set, 253–255
Bulk type bushing (*see* Ancillary equipment)

Capacitance
grading, 190
line, 176
tap, 190
winding, 178
Carbon dioxide, 262
Carbon monoxide, 262
Cellulose, 262
affinity for moisture, 266
Characteristic admittance, 178–179
Chi-squared distribution, 269–271
Chopped wave, 180–181, 256
Coil (*see* Winding)
Cold trap, 268
Confidence limit, 269
Connection diagram (*see* Nameplate information)
Cooling class, 99–102, 197–199
nameplate, 222–223
Cooling equipment (*see* Ancillary equipment)
Core
butt lap, 17
circular, 18, 23
construction, 16–19
design, 16–19, 22–23
laminations, 16–19, 23, 77–79
losses, 13–14, 32, 58
mitered, 17–18, 77
volume, 13
Core form design, 18, 22–23
five-legged, 58, 60
zero-phase sequence impedance, 86–92
three-legged, 56–58
zero-phase sequence impedance, 86–92
Core ground, 79
Core ground test (*see* Testing)
Corona ring, 191
Current
circulating, 16, 94–95, 139–140
conventional, 3
eddy, 74–75, 77
electron, 3
excitation, 242, 254–255
fault (*see* Faults)
harmonics (*see also* Third harmonic current), 16
in delta connection, 33–35
in three-legged core, 57–58
in wye-wye connection, 27–28, 30
in zigzag connection, 44
resonance with, 32
induced, 58
inrush (*see* Exciting current inrush)
leakage, 258

[Current]
magnetizing, 7, 10, 15–16, 27, 32
primary, 21–22
secondary, 21–22
tank heating, 169–172
zero-phase sequence, 27, 29, 169–170
Cutoff frequency, 179

DC offset, 150
Delta connection (*see* Two-winding connections)
Dew point, 267
Draw lead bushing (*see* Ancillary equipment)
Drying, 268

Equivalent circuit
positive phase sequence, 81
three-winding, 136–138
two-winding, 80–81
zero phase sequence, 85–87
Ethane, 262, 264
Ethylene, 262, 264
Exciting current inrush, 165–169
Extended warranty, 274–275

Failure rate, 269
Faults (*see also* Short circuits)
duration, 148–149
limiting, 148
phase sequence diagrams, 84
phase-to ground, 109–111, 151
phase-to-phase, 153
phase-to-phase-to-ground, 152
thermal, 264
three-phase, 93, 150
Ferroresonance, 32, 175–177
Field intensity, 3–4, 6, 9
Filter press, 268
Flux, 2–3, 7
density, 3, 76–77, 155–156
excluded, 72
leakage, 69–74, 92, 138, 155

[Flux]
path, 58–61
residual, 167–168
stray, 58, 147–148
zero-phase sequence, 88–89, 169
Forced air cooling (*see* Ancillary equipment)
Full wave, 180–181, 256
Fuse operations, 172–175, 186–187
tank overheating from, 170

Gauss, 2
Georg machine, 16
Gilbert, 2
Ground current isolation, 35, 44
Grounding bank, 35, 40–42, 86–87

Handling and shipping, 236–237
Heat run test, 101, 257
Hydrocarbon molecules, 261–262
Hydrogen, 261
Hysteresis, 11–13

Ideal transformer, 80, 136
Impedance
autotransformer, 122–128
matching, 93–98
open-circuit, 85–88
series, 80
autotransformer, 122–128
regulation and, 92–94
three-winding, 137–138
zero-phase sequence, 85–89
zero-phase sequence, 85–92, 112
Inductance, 8, 167–168, 178
Inrush current (*see* Exciting current inrush)
Inspection
acceptance, 237–238
internal, 238–239
Insulating fluid (*see* Oil)
Insulation
basic impulse level, 182, 232
bushing, 189–191

[Insulation]
composition of, 262
decomposition of, 262, 268
high performance, 226
moisture (*see* Moisture)
power factor, 257–258
resistance, 241, 257
testing (*see* Testing)
turn-to-turn, 177–178, 180–182
Interconnected star (*see* Two-winding connections)

Key gases, 262

Laplace transform, 178–180
Lightning arrestor, 182
Lightning impulse test (*see* Testing)
Lightning-induce failures, 180–182
Line drop compensation, 228
Load balancing scheme, 211–212, 215–218
Load tap changer (*see* Ancillary equipment)
Loss of life (*see* Thermal capability)
Losses
conductor, 74–76
test report, 100–101
transposed conductor, 75–76
copper, 75, 101
core, 13–14
five-legged, 58
overexcitation, 32
reducing, 76–79
test report, 100–101
iron, 76
no-load, 76
testing (*see* Testing)
Low-voltage impulse, 258

Magnetic
centers, 159–161, 185–186, 203–204
circuit, 6–10
core, 5–6
[Magnetic]
field intensity (*see* Field intensity)
flux (*see* Flux)
induction, 4
materials, 6, 10–11, 16–19
units and conversion factors, 1–2
Magnetomotive force, 2, 21–22, 135
short circuit, 159
Magnetostriction, 20
Maintenance
checklist, 246–251
predictive, 252, 261
preventative, 252
spare transformer, 245
Maxwell, 2
Megger test (*see* Testing)
Methane, 262, 263
Moisture
in insulation, 266
oil analysis, 265–267
removal of, 268

Nameplate information, 221–233
basic impulse level, 232
cooling class, 223
impedance, 231–232
KVA rating, 225–226
layout, 233
vector diagram, 229
voltage rating, 224–225
weights, 229
winding connection diagram, 226–229
Neutral
common, 26
grounded, 30, 33
open circuited, 27–29
zero-sequence current path, 33

Oersted, 2
Oil
capacity, 229–230
coking, 204
composition of, 261–262

[Oil]
contamination, 192–193, 195, 268
cracking, 262
decomposition, 261–262
filling (*see* Vacuum oil filling)
moisture (*see* Moisture)
oxidation, 194–195
preservation (*see* Ancillary equipment)
testing (*see* Testing)
Overexcitation, 32
Overheating
insulation, 262
oil, 262
overload (*see* Thermal capability)
short circuit, 147–148
stray flux, 58, 147–148
tank current, 32, 169–172

Parallel operation, 93–96
with voltage regulating transformers, 210–212
Partial discharge test (*see* Testing)
Permeability, 6, 17, 72
Phase displacement, 36–40
autotransformer connection, 129
delta-wye connection, 65
wye-wye connection, 27
zigzag connection, 44
Phase rotation, 26, 36–40, 65
Phase shifting transformer, 212–215
Piper chart, 266
Poisson distribution, 271
Polarity, 62–64
Pressure gauge (*see* Ancillary equipment)
Preventative autotransformer, 205
Primary conductor, 5, 20
Probability
chi-squared, 269–271
failure, 271–272
Poisson, 271–272
unavailability, 272–273
Propagation constant, 178–179

Radiator (*see* Ancillary equipment)
Rating
basic impulse level, 182
frequency, 99–100
KVA and MVA, 98–99
nameplate, 221–222, 225–226
temperature (*see* Thermal capability)
voltage, 99
nameplate, 224
Reactance
leakage, 31, 74, 80, 92
magnetizing, 31–32, 79–80, 85
resonance with line capacitance, 175
Real power, 53
Regulation, 92–93
Reluctance, 7
Rogers ratio method, 262–264

Saturation, 10–11
autotransformer short circuit, 130
core size effect on, 45
exciting current inrush, 165–67
magnetizing reactance effect on, 79, 95
shell-form core, 61
zero-phase sequence excitation, 88–89
Secondary conductor, 5, 20
Sequence networks, 82–84
Shell form design, 59–62
short-circuit forces, 162–163
zero-phase sequence impedance, 86, 88
Short circuit (*see also* Faults), 147–155
asymmetrical peak current, 149, 154
DC offset, 150
forces, 155–165, 185–186
heating effects, 147
stray flux, 58, 147–148
symmetrical current, 149
testing (*see* Testing)
winding strength, 158–165

Skin effect, 75
Static charge, 238, 245
Statistical economics, 273–275
Steel
  amorphous, 77
  burrs, 77
  grain-oriented, 6, 17–18, 76–77
  NiFe, 6
  oxide film, 77–80
  permaloy, 6
  SiFe, 6
  trade-off with copper 47–48
Storage, 237
Surge arrestor (*see* Lightning arrestor)
Switching impulse test (*see* Testing)
Symmetrical components
  diagrams, 84
  method, 81–85
  transformations, 83
Symmetrical short-circuit current, 149
Sympathetic inrush, 169

Tap changer (*see* Ancillary equipment)
Temperature (*see also* Thermal Capability)
  ambient, 102
  average winding, 100–103
  calculation of, 102–105, 107–108, 114
  hot-spot winding, 102–103
  top oil, 100, 102–103
Temperature rise test (*see* Heat run test)
Terminal markings, 62–65
Tesla, 2
Test report, 99–102
Testing
  applied voltage, 256
  carbon monoxide, 238
  combustible gas, 238
  core ground, 239–240
  dielectric, 256
    oil, 268
[Testing]
  Doble, 240, 258
  excitation current, 242, 254–255
  factory, 252, 255–258
  field, 258–261
  gas chromatograhy, 262
  induced voltage, 256
  insulation resistance, 257
  lightning impulse, 256
  load losses, 255
  megger, 240–241, 257–258
  no-load losses, 255
  oil, 241–242
    dielectric, 268
    dissolved combustible gas, 261–265
    moisture, 265–267
  oxygen, 238, 242
  partial discharge, 256
  switching impulse, 256
  turns ratio (TTR), 240, 252–255, 258–261
  water-in-oil analysis, 265–267
Thermal capability, 101
  aging acceleration factor, 105–106, 108
  derating from loss of cooling, 218–220
  loss of life, 105–106
  three-winding transformers, 141
Third harmonic current
  magnetizing current, 16
  suppression, 33–35
  three-legged core form 57–58
  wye-wye connection, 30, 32
  zigzag connection, 44
Three-winding connections, 133–146
  construction of, 138
  nameplate 226
  short-circuit forces, 163–165
  tertiary winding (*see* Winding)
  test report of, 144–145
  transformer laws, 134–135
Total combustible gas, 263

Transformer laws
  autotransformer, 119
  two-winding, 21–22
Transformer turns ratio, 21
  matching, 93–95
TTR test (*see* Testing)
Two-phase transformations, 50–56
Two-winding connections, 25–67
  comparisons of economy, 44–47
  delta-delta, 90
  delta-wye, 33–40, 89–90
  interconnected star, 42 (*see also* zigzag)
  open delta, 49–50
  Scott, 51–56
  tee, 50–51, 111–112
  wye-delta, 33–42, 89–90
  wye-wye, 26–33, 88–92
  zigzag, 42–44, 261–261

Unbalanced conditions (*see* Symmetrical components)

Vacuum drying, 268
Vacuum oil filling, 230, 242–245
Vapor pressure, 267–268
Voltage
  applied, 21–22, 70–72
    testing (*see* Testing)
  breakdown, 268
  collapse, 32
  excitation, 15, 77
  extinction, 256
  inception, 256
[Voltage]
  induced, 4–6
    in core, 16
    in primary winding, 72
    in secondary winding, 20–21
    testing (*see* Testing)
    volts per turn, 11–12
  non-sinusoidal, 28
  root mean square, 14
  sinusoidal, 27–28, 30, 33
  surges, 177–183
Voltage regulating transformer, 209–210

Weber, 2
Winding
  clamping, 162
  conductor, 74–76
  disc-type, 161–162
  forces (*see* Short circuit)
  pancake, 59
  placement of, 20–23
  primary, 20, 138
  secondary, 20, 138
  short-circuited (*see* Short circuit)
  tertiary, 57, 133–134, 138
    imbedded, 134
    stabilizing effect of, 141–144
Winding hot spot, 99
Wye connection (*see* Two-winding connections)

Zigzag connection (*see* Two-winding connections)

Printed in the United States
128987LV00002B/123/A

9 780824 707668